# Nanomedicine and Nanotoxicology

**Series editor**

V. Zucolotto, São Carlos, Brazil

More information about this series at http://www.springer.com/series/10620

Hao Wang · Li-Li Li

Editors

# In Vivo Self-Assembly Nanotechnology for Biomedical Applications

 Springer

*Editors*
Hao Wang
CAS Center for Excellence in Nanoscience,
  CAS Key Laboratory for Biomedical
  Effects of Nanomaterials and Nanosafety
National Center for Nanoscience and
  Technology (NCNST)
Beijing
China

Li-Li Li
CAS Center for Excellence in Nanoscience,
  CAS Key Laboratory for Biomedical
  Effects of Nanomaterials and Nanosafety
National Center for Nanoscience and
  Technology (NCNST)
Beijing
China

ISSN 2194-0452          ISSN 2194-0460   (electronic)
Nanomedicine and Nanotoxicology
ISBN 978-981-13-3900-4          ISBN 978-981-10-6913-0   (eBook)
https://doi.org/10.1007/978-981-10-6913-0

Printed on acid-free paper

This Springer imprint is published by the registered company Springer Nature Singapore Pte Ltd.
part of Springer Nature
The registered company address is: 152 Beach Road, #21-01/04 Gateway East, Singapore 189721, Singapore

# Foreword

With the rapid development of nanoscience and technology in the past decades, nanotechnology has become a multidisciplinary and innovative field and extensively merged with various fields, such as physics, chemistry, biology, medicine.

Among them, nanomaterials, nanomedicine, and nanobiotechnology gradually strengthen and develop as emerging fields because they significantly affect our public healthcare systems and people's daily life. Therefore, we envision that the application of nanotechnology will lead a revolutionary paradigm in disease diagnostics and therapeutics.

In this book, we review the recent advances in nanomaterials, nanotechnology, and their applications in biomedical field. In particular, the self-assembled nanomaterials are systemically discussed since their meaningful fundamental science and potential clinical translation. Inspired by nature, the self-assembly in organism plays a significant role in fabricating biological entities and regulating life processes.

Interestingly, the biofunctionality in life relies on the self-assembled structures and in most case the diverse biological function appears upon the self-assembled structures that grow to nanoscaled size. Hence, people keep attempting to understanding and mimic the natural nanosized self-assembly. Beyond nature heuristic, artificially designed self-assembled nanomaterials also obtained the considerable development. Self-assembled nanomaterials can be modulated by intelligent stimuli due to the intrinsic dynamic property.

With the development of the synthetic chemistry and material science, the variable molecules can be considered as building blocks for constructing nanomaterials. The endless imagination drives the design of the novel self-assembled nanomaterials with well-controlled nanostructures and fantastic biofunctions. So far, the method of preparation of nanomaterials with defined parameters such as size, shape, charge, payload, and surface chemistry has been widely investigated, providing the necessary support for reliable application and translation. However, when we utilize the sophisticated nanomaterials in biological environment, their function somewhat compromises upon interaction of complexed biomolecules or biointerfaces.

The concept of in vivo self-assembly is newly developed by us. Generally, by introducing exogenous molecules, they in situ simultaneously self-assembled into well-ordered structures in living objects and reveal biological functions. The remarkable keypoint of this concept is that the self-assembly does not initialize outside the living objects, but "smart" self-assembled in a certain region in vivo. Why do we need in vivo self-assembly? Compared to small molecules, nanomaterials exhibit excellent behavior in biomedical diagnostics and therapeutics, while the small molecules are superior to simplicity and reliability. By leveraging the advantages of small molecules and nanomaterials, in vivo self-assembly strategy represents huge potential in fundamentals and translations. The unique assembly-/aggregation-induced retention (AIR) effect of in vivo self-assembly greatly improves the performance of diagnostics and therapeutics.

Given the fact that the emergence of in vivo self-assembly nanotechnology not only in its intrinsic scientific merits but also in healthcare benefit, I truly believe that this book provides the readers with a new sight in nanotechnology and may represent a useful reference for chemists, material scientists, biologists, and medical doctors.

Beijing, China                                                                    Prof. Dr. Yuliang Zhao
January 2018                                      CAS Center for Excellence in Nanoscience,
                                                        CAS Key Laboratory for Biomedical Effects
                                                                    of Nanomaterials and Nanosafety,
                                                National Center for Nanoscience and Technology
                                                                                                      (NCNST)

# Contents

# Contributors

**Hong-Wei An** CAS Center for Excellence in Nanoscience, CAS Key Laboratory for Biomedical Effects of Nanomaterials and Nanosafety, National Center for Nanoscience and Technology (NCNST), Beijing, China; Institute of High Energy Physics, Chinese Academy of Sciences (CAS), Beijing, China

**Yu-Juan Gao** CAS Center for Excellence in Nanoscience, CAS Key Laboratory for Biomedical Effects of Nanomaterials and Nanosafety, National Center for Nanoscience and Technology (NCNST), Beijing, China

**Yuan Gao** Beijing Engineering Research Center for Bionanotechnology and CAS Key Laboratory for Biomedical Effects of Nanomaterials and Nanosafety, CAS Center for Excellence in Nanoscience, National Center for Nanoscience and Technology, Beijing, People's Republic of China

**Zhentao Huang** Beijing Engineering Research Center for Bionanotechnology and CAS Key Laboratory for Biomedical Effects of Nanomaterials and Nanosafety, CAS Center for Excellence in Nanoscience, National Center for Nanoscience and Technology, Beijing, People's Republic of China

**Li-Li Li** CAS Center for Excellence in Nanoscience, CAS Key Laboratory for Biomedical Effects of Nanomaterials and Nanosafety, National Center for Nanoscience and Technology (NCNST), Beijing, China

**Yao-Xin Lin** CAS Center for Excellence in Nanoscience, CAS Key Laboratory for Biomedical Effects of Nanomaterials and Nanosafety, National Center for Nanoscience and Technology (NCNST), Beijing, China; University of Chinese Academy of Sciences (UCAS), Beijing, People's Republic of China

**Guo-Bin Qi** CAS Center for Excellence in Nanoscience, CAS Key Laboratory for Biomedical Effects of Nanomaterials and Nanosafety, National Center for Nanoscience and Technology (NCNST), Beijing, China

**Zeng-Ying Qiao** CAS Center for Excellence in Nanoscience, CAS Key Laboratory for Biomedical Effects of Nanomaterials and Nanosafety, National Center for Nanoscience and Technology (NCNST), Beijing, China

**Sheng-Lin Qiao** University of Chinese Academy of Sciences (UCAS), Beijing, People's Republic of China; CAS Center for Excellence in Nanoscience, CAS Key Laboratory for Biomedical Effects of Nanomaterials and Nanosafety, National Center for Nanoscience and Technology (NCNST), Beijing, China

**Hao Wang** University of Chinese Academy of Sciences (UCAS), Beijing, People's Republic of China; CAS Center for Excellence in Nanoscience, CAS Key Laboratory for Biomedical Effects of Nanomaterials and Nanosafety, National Center for Nanoscience and Technology (NCNST), Beijing, China

**Lei Wang** CAS Center for Excellence in Nanoscience, CAS Key Laboratory for Biomedical Effects of Nanomaterials and Nanosafety, National Center for Nanoscience and Technology (NCNST), Beijing, China

**Man-Di Wang** Institute of High Energy Physics, Chinese Academy of Sciences (CAS), Beijing, China

**Yi Wang** CAS Center for Excellence in Nanoscience, CAS Key Laboratory for Biomedical Effects of Nanomaterials and Nanosafety, National Center for Nanoscience and Technology (NCNST), Beijing, People's Republic of China; University of Chinese Academy of Sciences, Beijing, People's Republic of China

# Chapter 1
# Supramolecular Self-assembled Nanomaterials for Fluorescence Bioimaging

**Lei Wang and Guo-Bin Qi**

**Abstract** In recent years, the construction of self-assembled nanomaterials with diversified structures and functionalities via fine tune of supramolecular building blocks increases rapidly. The self-assembled nanomaterials are potential high-efficient probes/contrast agents for high-performance fluorescence imaging techniques, which show distinguished advantages in terms of the availability of biocompatible imaging agents, maneuverable instruments, and high temporal resolution with good sensitivity. Generally, the self-assembled nanomaterials show high stability in vivo, prolonged half-life, and desirable targeting properties. Furthermore, self-assembled nanomaterials can be modulated by intelligent stimuli due to the dynamic nature of supramolecular materials. In this chapter, we will summarize recent advances in smart self-assembled nanomaterials as fluorescence probes/contrast agents for biomedical imaging, including long-term imaging, enzyme activity detection, and aggregation-induced retention (AIR) effect for diagnosis and therapy. The self-assembled nanomaterials will be classified into two groups, i.e., the ex situ and in situ self-assembled nanomaterials based on the assembly mode and the strategy for fluorescence regulation. Finally, we conclude with an outlook toward future developments of self-assembled nanomaterials for fluorescence bioimaging.

**Keywords** Supramolecular · Nanomaterials · Fluorescence bioimaging
In situ self-assembly

L. Wang (✉) · G.-B. Qi
CAS Center for Excellence in Nanoscience, CAS Key Laboratory for Biomedical Effects of Nanomaterials and Nanosafety, National Center for Nanoscience and Technology (NCNST), No. 11 Beiyitiao, Zhongguancun, Haidian District, Beijing 100190, China
e-mail: wanglei@nanoctr.cn

G.-B. Qi
e-mail: qigb@nanoctr.cn

© Springer Nature Singapore Pte Ltd. 2018
H. Wang and L.-L. Li (eds.), *In Vivo Self-Assembly Nanotechnology for Biomedical Applications*, Nanomedicine and Nanotoxicology,
https://doi.org/10.1007/978-981-10-6913-0_1

1

## 1.1 Introduction

Inspired by sophisticated biological structures and their physiological processes, supramolecular chemistry inherits the spontaneous self-assembly of building blocks from their natural counterparts, which is driven by a variety of noncovalent interactions. Since the Nobel Prize was awarded to Lehn, Cram, and Pedersen for their pioneering works on guest-host and supramolecular chemistry in 1987, [1] supramolecular material science has undergone significant expansion. A large amount of materials self-assembled from functional building blocks through noncovalent interactions, such as hydrogen bonding, $\pi-\pi$, hydrophobic interactions, were designed and prepared, the structures and properties of which could be regulated by internal/external stimuli [2, 3]. Meanwhile, by emerging the modern nanoscience and biotechnology, a new research field of supramolecular nanomaterials has been gradually developed and attracted wide attention for biomedicine. Besides the traditional pre-organized nanomaterials prepared in solution, the in situ construction of self-assembled nanomaterials in vivo becomes a promising research field facing the structure evolution of self-assembled nanomaterials at biointerfaces.

Fluorescence is the emission of light from any substance and occurs from electronically excited states. In excited singlet states, the electron in the excited orbital is paired (by opposite spin) to the second electron in the ground-state orbital. Consequently, return to the ground state is spin allowed and occurs rapidly by emission of a photon. Based on the fluorescence phenomena and mechanism, fluorescence imaging technique enables highly sensitive, less-invasive, cheap, and safe detection using readily available instruments without requirements for the expense and difficulty of handling radioactive tracers for most biochemical measurements [4, 5]. To achieve desired signal output and high signal-to-noise ratio, a variety of fluorescent nanoprobes/contrast agents have been explored and attracted increasing attention for fluorescence imaging. To date, various fluorescent probes/contrast agents based on organic/polymeric nanomaterials have been developed for in vitro/in vivo imaging, which show negligible toxicity and superior capacity to combine multiple modalities on one nanomaterial. Furthermore, supramolecular self-assembled nanomaterials show outstanding advantages of high sensitivity and deep insight for in vivo processes [6, 7]. In this chapter, the fluorescent self-assembled nanomaterials were divided into ex situ construction in solution and in situ construction in vivo (Fig. 1.1). For ex situ nanoassemblies, great efforts should be paid for the preparation of defined superstructures and functionalities with specific supramolecular interactions. For in situ construction of self-assembled nanomaterials in living systems, we will focus on sophisticated responsiveness under physiological/pathological conditions to accomplish diagnosis and/or therapy of diseases in vitro and in vivo.

Generally, the isolated organic molecules in dilute solution are utilized to measure the fluorescence without intermolecular interactions. However, for most organic dyes, many of them in the supramolecular nanomaterials as nanoprobes/contrast agents are actually in the condensed state as well as a reduction in

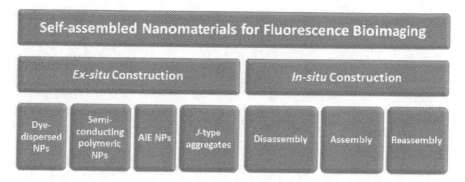

**Fig. 1.1** Different classes of self-assembled nanomaterials for fluorescence bioimaging

fluorescence by aggregation-caused quenching (ACQ), [8] which is different from that in dilute solution. That is, supramolecular fluorescence nanomaterials usually show weakened or quenched fluorescence due to the ACQ effect, where the significant collisional quenching occurs between the fluorescence molecules. To address this issue, the supramolecular fluorescence nanomaterials should be designed and prepared by dispersing organic dyes down to a level avoiding concentration quenching in the nanomatrix or introducing bulky groups to prevent the aggregation [9, 10]. Recently, several novel systems have been developed to produce highly bright supramolecular fluorescence nanomaterials. One is the specialized aggregation-induced emission (AIE) and *J*-type aggregates. These fluorescence molecules do not emit in dilute solution, but they give off bright fluorescence in the condensed state. Another is dye-doped semiconducting polymer nanoparticles with high fluorescence brightness [11]. The highly fluorescent nanosized supramolecular nanoprobes/contrast agents were prepared by incorporating with functional polymer matrix, which could be further modified for targeted imaging and diagnostics. In this chapter, we will summarize recent advances in smart self-assembled nanomaterials as fluorescence probes/contrast agents for biomedical imaging, including long-term imaging, enzyme activity detection, and aggregation-induced retention (AIR) effect for diagnosis and therapy. The self-assembled nanomaterials will be classified into two groups, i.e., the ex situ and in situ self-assembled nanomaterials based on the assembly mode and the regulation strategy of fluorescence. Finally, we conclude with an outlook toward future developments of self-assembled nanomaterials for fluorescence bioimaging.

## 1.2 Ex-Situ Construction

### 1.2.1 Dye-Dispersed Nanoparticles

To prevent the ACQ effect, organic dyes can be encapsulated into polymeric matrix to form promising nanoprobes/contrast agents for bioimaging. Furthermore, dye-dispersed nanoparticles could efficiently avoid photobleaching and produce amplified fluorescence signals due to the incorporation of several dye molecules.

Generally, hydrophobic fluorescent dyes are encapsulated within hydrophobic core of amphiphilic polymeric matrix nanoparticles (NPs) through hydrophobic interactions, which prevents the dyes aggregation and increases the stability and compatibility. The supramolecular strategies were widely used to prepare nano-materials as nanoprobes/contrast agents for bioimaging in vitro and in vivo. Generally, the amphiphilic polymers, such as poly(acrylic acid) (PAA), [12] poly (lactic-co-glycolic acid) (PLGA), [13] and polyurethane, [14] self-assembled into nanoparticles by precipitation method with hydrophobic core and hydrophilic outer layer. The nanoparticles as nanomatrix were loaded with small-molecule fluor-ophores to yield dye-dispersed polymeric NPs. Compared with small fluorescence dye, dye-dispersed NPs could produce a highly amplified optical signal due to the incorporation of many dyes. However, the quantity and distribution of organic dyes in the polymeric nanomatrix are the most challenge for effectively avoiding quenching.

It is difficult to accurately control the quantification and distribution of dispersed dyes in self-assembled nanoparticles. Therefore, the highly ordered structure such as micelles [15], nanofibers [16], nanowires [17], and nanotubes [18] are con-structed through defined supramolecular interactions, where small organic dyes could be well loaded and controlled with desirable aggregation level and optical properties. For example, Wang and Frank prepared NIR absorbing and emission squaraine (SQ) dyes, which have good stability and high photoresistance. They modulated the aggregation of SQ dyes in amphiphilic phospholipid bilayers of liposomes to achieve NIR fluorescence and PA tomography dual-modular imaging (Fig. 1.2) [19]. The liposomes with variable mixing ratios between SQ dyes and the liposome are typical vesicular structures with a size of $76 \pm 15$ nm, which is in accordance with the dynamic light scattering (DLS) result (Size: $103 \pm 14$ nm, PDI: 0.18). When doping minimal amounts of SQ (1:500), molecularly dispersed SQ in bilayers shows remarkable fluorescence (Fig. 1.2). Interestingly, the PA signals are enhanced with the increase of SQ dyes in the nanoconfined bilayer region, which are due to the formation of SQ-based H-aggregates and enhanced thermal conversion efficiency (Fig. 1.2). NIR fluorescence imaging in vivo indi-cated that the majority of SQ $\subset$ L are enriched in the area where the blood vessels are generated. The research indicated that the importance of supramolecular mod-ulation (monomer or aggregation) of NIR dyes loaded in the nanomatrix, which will determine the characteristics and application of the dye-dispersed nanomaterials. For fluorescence bioimaging, the dyes should be well dispersed in monomer states to prevent the ACQ effect and obtain high fluorescence signals.

Besides construction of nanomatrix with highly ordered structure, doping with ionic liquid-like salts of a cationic dye with a bulky hydrophobic counterion that serves as spacer is another strategy to prevent undesirable aggregation for dye-dispersed nanoparticles. In this case, the polymeric nanoparticles can be loaded with up to 500 dyes and show higher brightness than quantum dots due to the minimization of dye aggregation and self-quenching, which also exhibit

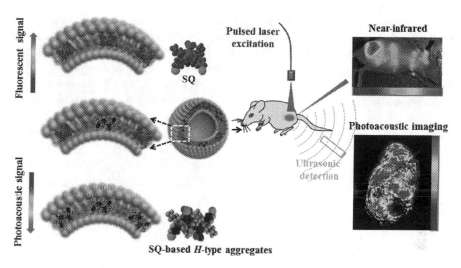

**Fig. 1.2** Schematic diagram of the nanoconfined SQ assemblies in phospholipid bilayer and the utilization of these vesicles as probes for dual-modular tumor imaging in vivo. (Reproduced with permission. Copyright 2014, American Chemical Society)

photoinduced reversible on/off fluorescence switching. These dye-dispersed NPs show absence of toxicity and can be utilized for super-resolution cell imaging with high signal-to-noise ratio. Moreover, the fluorescence obtained from chemiluminescence was applied to image hydrogen peroxide, which is overexpressed under specific pathological condition. Lee et al. demonstrated that nanoparticles formulated from peroxalate esters and fluorescent dyes can image hydrogen peroxide in vivo with high specificity and sensitivity. The peroxalate nanoparticles can be loaded with fluorescence dyes, where chemiluminescent reaction between peroxalate esters and hydrogen peroxide produce light from fluorescent dyes. The emission wavelength of the peroxalate nanoparticles can be tuned by incorporating different dyes from 460 to 630 nm. Furthermore, excellent sensitivity and specificity for hydrogen peroxide over other reactive oxygen species are highly needed for in vivo imaging. The peroxalate nanoparticles are demonstrated with the capability of imaging hydrogen peroxide in the peritoneal cavity of mice during a lipopolysaccharide-induced inflammatory response.

Proteins, such as human serum albumin (HSA) and bovine serum albumin (BSA), are of the nanoscale in size with excellent biocompatibility. They can act as nanocarrier to combine with fluorescent molecules via self-assembly, forming fluorescent nanoprobes/contrast agents. HSA is a good biological nanosized carrier, which is used to self-assemble complexes with the NIR dye [20–23]. The self-assembled nanoparticles with high fluorescence quantum yield can be utilized for NIR bioimaging. On the other hand, they showed photothermal effect or photodynamic properties due to the NIR dyes, which can further show therapeutic effect for cancer and so on. BSA is another protein-based nanomatrix, which can carrier organic dyes through hydrophobic interactions to form fluorescence nanoprobes/contrast agents [24]. SQ dyes are a kind of promising NIR fluorophores with high absorption coefficients, bright fluorescence, and photostability. Wang and Frank utilized BSA as a biocompatible carrier to obtain a highly stable supramolecular adducts of SQ and BSA (SQ ⊂ BSA) for tumor-targeted imaging and photothermal therapy in vivo (Fig. 1.3). SQ was selectively bound to BSA hydrophobic domain via hydrophobic and hydrogen-bonding interactions with up to 80-fold enhanced fluorescence intensity (Fig. 1.3a, b). By covalently conjugating target ligands to BSA, the SQ ⊂ BSA is capable of targeting tumor sites. According to the biodistribution of SQ ⊂ BSA monitored by fluorescence imaging, the photothermal therapy can be carried out effectively in vivo (Fig. 1.3d). The intrinsic chemical instability and self-aggregation properties of NIR dyes in physiological condition could be solved by protein carrier.

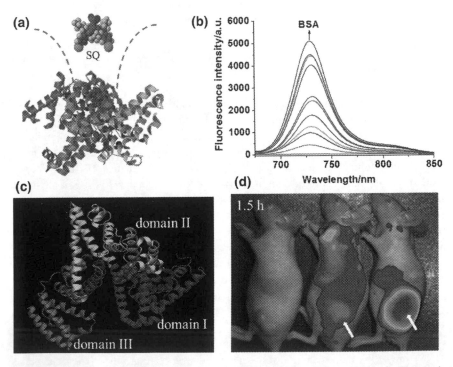

**Fig. 1.3 a** Schematic diagram of SQ dyes and BSA adducts (SQ ⊂ BSA). **b** Change in fluorescence intensity of SQ (1 μM) with the addition of BSA in phosphate buffer. The concentration of BSA was from 0, 1.1 to 12.1 μM. Excitation wavelength = 710 nm. **c** Molecular structure of site-selective binding of SQ with BSA: site I in domain II. **d** In vivo NIR optical imaging of nude mice bearing KB tumors at 1.5 after tail vein injection of the SQ ⊂ BSA and SQ ⊂ BSA-FA adducts. (The arrows show the location of subcutaneous tumors)

## 1.2.2 Semiconducting Polymer Nanoparticles

Semiconducting polymer nanoparticles (SPNs) emerge as a new class of promising fluorescence probes/contrast agents with excellent optical properties [25–29]. The advantages of SPNs include simple synthesis and separation procedures, excellent photostability, low cytotoxicity, high quantum yield. (Fig. 1.4) [30, 31]. The semiconducting polymer nanoparticles have been demonstrated for fluorescence, chemiluminescence, and persistent luminescence for real-time and targeted imaging in vitro and in vivo [30–32]. Generally, the SPNs are prepared by the reprecipitation method. The semiconducting polymer dissolved in a "good" solvent is rapidly added to an excess of "poor" solvent under ultrasonic dispersion. Later on, Chiu and his coworkers developed a modified method, which is called as nanoprecipitation method to obtain small and uniform SPNs with specific groups on the surface. Typically, an amphiphilic comb-like polystyrene polymer (PS-PEG-COOH)

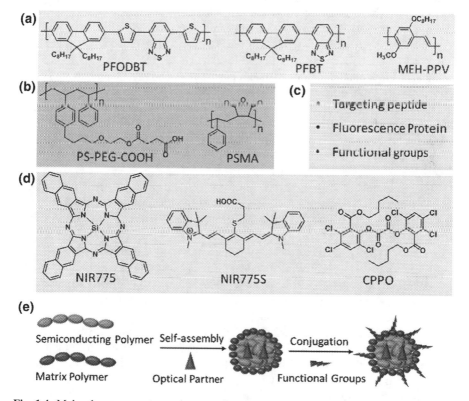

**Fig. 1.4** Molecular components and preparation process of SPNs. **a** The host semiconducting polymer PFODBT, PFBT, and MEH-PPV; **b** the matrix polymer including a PEG-grafted poly (styrene) copolymer and PSMA; **c** the conjugated functional groups; **d** the optical partner, such as FRET acceptor NIR775, ROS responsive dye NIR 755S, and chemiluminescent substrate CPPO that serves as CRET energy donor; **e** the schematic illustration of preparation process of SPNs

or poly(styrene-co-maleic anhydride) (PSMA) is used to functionalize highly fluorescent SPNs (Fig. 1.4a, b) [28, 29]. Firstly, a weight ratio of 4/1 of semi-conducting polymer to matrix was mixed in THF to form a homogeneous solution. Secondly, the resulting THF solution was added to two fold volume of water under ultrasonication. Finally, the organic solvent was evaporated under an inert gas atmosphere. The SPNs have an average diameter of about 15 nm and contain more than 80% of effective fluorophores. Moreover, this strategy produces surface-functionalized NPs with both PEG and carboxylic groups, which ensure good stability and high reactivity with biomolecules, such as antibodies or strep-tavidin. By using this strategy, they prepared highly stable SPNs, which approxi-mately 15 times brighter than the commercial quantum dots for in vivo tumor imaging. Similarly, they attached the temperature sensitive dye Rhodamine B (RhB), whose emission intensity decreases with increasing temperature within the matrix of SPNs [26]. Accordingly, they employed SPN-RhB as a ratiometric temperature sensor for measuring intracellular temperatures in a live-cell imaging mode. The exceptional brightness of SPN-RhB allows this nanoscale temperature sensor to be used also as a fluorescent probe for cellular imaging.

Rao and his coworkers have been worked on SPNs for several years. They have demonstrated for fluorescence, chemiluminescence, and photoacoustic imaging in living cells and small animals (Fig. 1.4c–e) [11, 30–35]. They report the first example of SPNs that can emit persistent luminescence with a lifetime of nearly 1 h after a single excitation exposure to white light and show potential for in vivo optical imaging [32]. The SPNs show both NIR persistent luminescent and fluorescent emissions. The persistent luminescent emission has been applied for in vivo optical imaging in mice, which provides support for future applications in cell tracking and molecular imaging. Moreover, they introduce fluorescence reso-nance energy transfer (FRET) and chemiluminescence resonance energy transfer (CRET) mechanism into SPN system to detect drug-induced reactive oxygen species (ROS) and reactive nitrogen species (RNS), which can direct evaluate acute hepatotoxicity [31]. Briefly, the ROS and RNS responsive motif was loaded into the SPNs to form FRET system and CRET system with different emission wavelength, respectively. The responsive SPNs simultaneously and differentially detect RNS and ROS using two optically independent channels. They image ROS and RNS to reflect drug-induced hepatotoxicity and its remediation longitudinally in mice after systemic challenge with acetaminophen or isoniazid. Similarly, Rao and his coworkers designed self-luminescing NIR nanoparticles by integrating biolumi-nescence resonance energy transfer (BRET) and FRET mechanism [30]. The NIR SPNs were prepared by nanoprecipitation using poly[2-methoxy-5-((2-ethylhexyl) oxy)-p-phenylenevinylene] (MEH-PPV, FRET donor) and NIR775 (FRET accep-tor) as the emitters and PS-PEG-COOH as the matrix, followed by conjugation with a bioluminescent eight-mutation variant of luciferase (Luc8) and RGD peptide. Three emission peaks from MEH-PPV, Luc8, and NIR775 are located at 480, 594, and 778 nm, respectively, indicating efficient BRET from Luc8 to MEH-PPV, followed by FRET from MEH-PPV to NIR775. Upon intravenous administration of the SNPs into mice, strong NIR fluorescence signals were observed from the lymphatic networks, including neck lymph nodes, axillary lymph nodes, inguinal lymph nodes, and lateral thoracic lymph nodes. The self-luminescing feature provided excellent tumor-to-background ratio for imaging very small tumors (2–3 mm in diameter). The novel self-luminescing nanoparticles provided highly improved sensitivity through bioluminescence imaging in in vivo studies, such as lymph-node mapping and cancer imaging.

The SPNs were prepared through microemulsion method with, poly (DL-lactide-co-glycolide) (PLGA) as matrix polymer, which is a Food and Drug Administration approved biocompatible polymer [36]. The PLGA-encapsulated SPNs (240–270 nm) have the terminal carboxyl groups of PLGA, which can be further conjugated with functional groups before or after SPN formation. Meanwhile, Liu and coworkers developed a general strategy through a single emulsion method to fabricate a series of SPNs with PLGA as nanomatrix, which showed emission in the range of 400–700 nm [37]. The quantum yields of the SPNs were 4–25 folds lower than those of semiconducting polymer in THF solutions, due to extensive semiconducting polymer aggregation in nanoparticles. Liu et al. introduced bulky groups (POSS) to poly(fluorenevinylene) to prevent close-packing of semiconducting polymer inside the SPNs. The reformative poly(fluorenevinylene) and PLGA-based SPNs showed an improved quantum yield of 19% relative to 11% for poly(fluorenevinylene) [38]. The surface carboxyl groups of PLGA nanoparticles were further modified with trastuzumab through coupling reaction. The trastuzumab-functionalized SPNs were then successfully used for specific discrimination of SK-BR-3 cells from MCF-7 and NIH/3T3 cells.

## 1.2.3 AIE Nanoparticles

Aggregation-induced emission (AIE) and aggregation-induced emission enhancement (AIEE) mechanism have been developed by Tang, which can obtain high fluorescence in aggregated state (Fig. 1.5a) [8, 39–47]. Tang and his coworkers discovered that AIE was exactly opposite to ACQ effect (Fig. 1.5b). The typical AIE fluorogenic molecules are nonemissive when molecularly dissolved but highly emissive when aggregated due to the restriction of intramolecular rotations (RIR). The AIE phenomenon is generally observed in molecules with rotating units such as phenyl rings. In dilute solutions, the rotor-containing fluorogens undergo low-frequency motions, resulting in fast nonradiative decay of the excited states and nonemissive of fluorogens. In the aggregates, the intermolecular steric interaction blocked the intramolecular rotations, leading to the emission of fluorogens when absorbed energy. The AIE effect made the fluorogens ideal candidate materials for the preparation of fluorescent nanoparticles with highly emissive fluorescent signals. In recent years, a large variety of AIE molecules have been developed, which have propeller-shaped structures and freely rotatable peripheral aromatic moieties. A few typical examples such as tetraphenylethene (TPE), hexaphenylsilole (HPS),

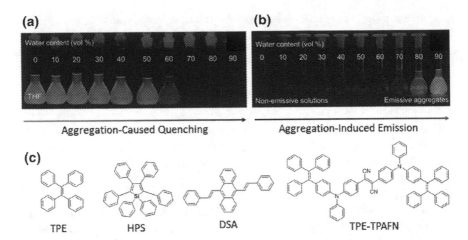

Fig. 1.5 **a** N,N-dicyclohexyl-1,7-dibromo-3,4,9,10-perylenetetracarboxylic diimide (DDPD) and **b** HPS in THF–water mixtures with different water contents for fluorescence photographs of solutions/suspensions. **c** Typical AIE luminogens TPE, HPS, DSA, TPE-TPAFN. (Reproduced with permission [39]. Copyright 2011, the Royal Society of Chemistry)

distyreneanthracene (DSA), and TPE conjugated 2,3-bis-[4-(diphenylamino)phenyl] fumaronitrile (TPE-TPAFN) are shown in Fig. 1.5c.

The AIE nanoparticles showed high stability and strong retention in living cells, which could be utilized for long-term cell imaging for many passages. For example, the aminated HPS derivatives spontaneously aggregated into ∼220 nm nanoparticles in aqueous media [48]. The nanoparticles effectively stained HeLa cells with bright green fluorescence and enabled long-term cell tracing up to four passages. Furthermore, the TPE based AIE nanoparticles may allow cell tracing for 10–15 passages [49, 50]. The strong retention of the AIE nanoparticles with high concentrations enabled visual monitoring of the cell growth, which was better than the commercial cell tracker.

BSA and functionalized BSA were also utilized to encapsulate an AIE molecule for bioimaging [51, 52]. For example, attaching of two TPE units with 2-2,6-bis [(E)-4-(diphenylamino)styryl]-4H-pyran-4-ylidene-malononitrile can afford an AIE molecule (TPE-TPA-DCM), which was imparted AIE characteristics by conjugation of AIE active TPE. The TPA-DCM in THF was added to BSA aqueous solution under sonication followed by crosslinking with glutaraldehyde, forming TPE-TPA-DCM-loaded BSA nanoparticles with bright fluorescence [51]. The TPE-TPA-DCM BSA nanoparticles can enter MCF-7 cells and accumulate in tumor for in vitro and in vivo imaging. Furthermore, the BSA was functionalized with arginine–glycine–aspartic acid (RGD) peptide to yield fluorescent probes for specific recognition of integrin receptor-overexpressed cancer cells.

Besides traditional conjugation of the targeting groups with AIE molecules to afford specific cell imaging, the functionalized AIE nanoparticles could be prepared by a precipitation strategy. The AIE molecules were encapsulated and stabilized with 1,2-distearoyl-sn-glycero-3-phosphoethanolamine-N-[methoxy (polyethylene glycol)] (DSPE-PEG) [53]. The functional groups such as targeting molecules and peptides can be introduced by co-assembly with pre-conjugated DSPE-PEG. The typical preparation process was shown in Fig. 1.6. THF solution of AIE molecule, DSPE-PEG$_{2000}$, and DSPE-PEG$_{5000}$-folate was added into water under continuous sonication, resulting in 90 nm AIE nanoparticles with fine-tuned surface folic acid density. The AIE nanoparticles could selectivity recognize MCF-7 from NIH3T3 cells due to the specific folate receptor-mediated endocytosis for MCF-7 cells. The supramolecular strategy showed many advantages including convenient preparation and well-controlled density of functional group on the surfaces.

The active DSPE-PEG$_{2000}$-R (R = NH$_2$, COOH, N-maleimide) could be also involved in the precipitation strategy to prepare AIE nanoparticles with surface functional groups [54–56]. As a result, the AIE nanoparticles could be post-modified with functional peptides, proteins and so on, which was much easier

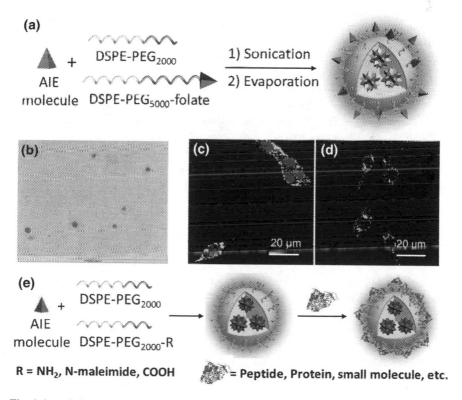

**Fig. 1.6 a** Schematic illustration of AIE nanoparticles preparation with DSPE-PEG as encapsulation matrix and the HR-TEM image (**b**). **c** One- and **d** two-photon confocal images of MCF-7 breast cancer cells after incubation with the nanoparticles. **e** Schematic illustration of the AIE-doped DSPE-PEG with post-modification. **a–d** were reproduced with permission from Ref [53]. Copyright 2011 Royal Society of Chemistry. **e** was reproduced with permission from Ref [8], Copyright 2013 American Society of Chemistry

than chemical modification of AIE molecule itself. For instance, DSPE-PEG$_{2000}$ and DSPE-PEG$_{2000}$-NH$_2$ were used to co-assemble with AIE molecules TPE-TPAFN. TPE-TPAFN-based AIE nanoparticles showed a uniform hydrodynamic size of 30 nm with NIR emission (maximum emission: 671 nm). AIE nanoparticles were conjugated with HIV-1 transactivator of transcription (Tat) via carbodiimide-mediated coupling (Fig. 1.6e) affords TPE-TPAFN-based DSPE-PEG-Tat nanoparticles. The Tat-AIE nanoparticles show higher stability and brightness than the commercial Qtracker® 655 and realize long-term noninvasive cell tracing. The Tat-AIE nanoparticles were capable of tracing MCF-7 cells for 10–12 generations in vitro. Moreover, in vivo fluorescence imaging through monitoring AIE nanoparticle-labeled C6 glioma cells in mouse indicated that the Tat-AIE nanoparticles had the ability to trace cell at long-term up to 21 days.

Near-infrared (NIR) AIE nanoparticles were highly desirable for bioimaging, especially for in vivo application as shown in the above examples, which display nearly no spectral overlap with biosubstrate autofluorescence. In another case, the shape-tailored organic quinoline–malononitrile (QM) with AIE characteristics is designed and synthesized to form AIE nanoprobes with far-red and NIR fluorescence and well-controlled morphology (from rod-like to spherical) [57]. The QM-based AIE nanoprobes are biocompatible and highly desirable for cell-tracking applications. The tumor-targeted bioimaging performance is related to the shape of QM nanoaggregates, where the spherical shape of QM nanoaggregates exhibits excellent tumor-targeted bioimaging in vivo. Similarly, diketo-pyrrolo-pyrrole (DPP) compounds with AIE properties are NIR emissive, which is stabilized with DSPE-PEG-Mal to form AIE nanoparticles. The surface of DPP-based AIE nanoparticles is further modified with cell penetrating peptide (CPP) to yield DPP-CPP nanoparticles with high brightness, good water dispersibility, and excellent biocompatibility for cell imaging [56].

Besides the NIR fluorescence imaging, the two-photon fluorescence imaging was developed for high-resolution imaging in vivo [58, 59]. Liu et al. reported the synthesis of ultrabright 4,7-bis[4-(1,2,2-triphenylvinyl)phenyl]benzo-2,1,3-thiadiazole (BTPEBT)-based supramolecular AIE nanoparticles with small size (≈33 nm), which showed a high two-photon absorption cross section, superb colloidal stability, excellent photostability, low in vivo toxicity and used for real-time intravital two-photon imaging of blood vessels [60]. The supramolecular AIE nanoparticles were brighter than widely used organic dyes such as Evans blue for two-photon blood vessels labeling agent. Furthermore, the supramolecular AIE nanoparticles can be used to visualize the major blood vessels as well as the smaller capillaries in the pia matter with a depth of beyond 400 μm.

## 1.2.4   J-Aggregates

J-type aggregates of fluorophores are another straightforward solution for ACQ effect. According to the supramolecular interactions, the H- and J-aggregates were developed with organic dyes. Based on the theoretical calculation and experimental results, H-aggregates showed blue-shifted absorption and quenched fluorescence due to the face-to-face dipole moments. However, the J-aggregates showed strong fluorescence with red-shifted absorption. Frank and Wang have been devoted to design supramolecular fluorescence based on perylenes by tuning the hydrogen bonds [61, 62]. Recently, Wang et al. designed and developed a new series of bis (pyrene) derivatives, i.e., BP1–BP4 with 1,3-dicarbonyl, pyridine-2,6-dicarbonyl, oxaloyl and benzene- 1,4-dicarbonyl as linkers, respectively (Fig. 1.7a) [63]. In solution, all compounds showed low fluorescence quantum yields ($\Phi < 1.7\%$) in variable organic solvents due to the twisted intramolecular charge transfer (TICT). In a sharp contrast, BP1 in the solid state was self-assembled to form J-type aggregates with almost 30-fold fluorescence enhancement ($\Phi$ was up to 32.6%) compared to that in solution (Fig. 1.7b, c). Nevertheless, H-type aggregates of BP3 and BP4 were observed with poor emissive efficiencies ($\Phi < 3.1\%$). Furthermore, the morphologies of supramolecular aggregates showed that BP1 and BP2 were dot-shaped nanoaggregates with 2–6 nm in diameters (Fig. 1.7d), while BP3 and BP4 showed sheet-like morphologies with 5–10 nm in width and 20–100 nm in length. The nanoaggregates of BP1 and BP2 coated with F108 surfactants showed good pH and photostability in physiological condition. Finally, the nanoaggregates of BP1 and BP2 were successfully employed as fluorescence nanoprobes for lysosome-targeted imaging in living cells with negligible cytotoxicity. To expand the application of BP derivatives, the functionalization of BP is achieved by introducing the active groups, such as COOH, $NH_2$, Br, $N_3$ (Fig. 1.7e).

To obtain the NIR emission under NIR two-photon irradiation, the BP1 was co-assembled with Nile red to form FRET two-photon fluorescence imaging

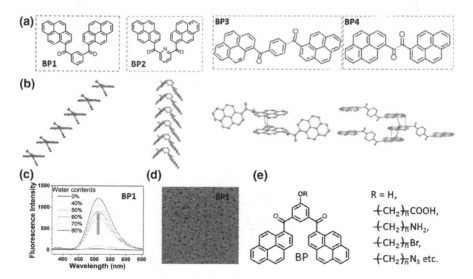

**Fig. 1.7** **a** Molecular structures of bis-pyrene derivatives BP1-4 and **b** their supramolecular packing structures. **c** Typical fluorescence spectra of BP1 upon aggregation and **d** the resulting TEM image. **e** Molecular structures of functionalized BP derivatives

contrast agents, which exhibited unprecedentedly high tissue penetration capability (Fig. 1.8) [64]. The BP-NR showed a large $\delta$ (2.4 × 10$^5$ GM) value, a NIR emission (630 nm), and a high energy transfer efficiency (60.7%). The $\delta$ value of this co-assembled BP-NR nanoparticle is the highest one among all the reported two-photon contrast agent materials. The BP-NR showed deep tissue penetration up to 2200 μm in-depth in mock tissue. Finally, the low-concentrated BP-NR (0.5 nM) was successfully applied to visualize the fine structures of aligned cartilage in the ear of mice in vivo.

Other typical fluorescence molecular probes can form J-aggregates after modification with π-conjugated groups due to tailored π−π stacking. For example, a series of lipidic boron-dipyrromethene (BODIPY) dyes were designed and synthesized to self-assemble into J-aggregates with NIR fluorescence [65]. The supramolecular J-type aggregates could act as imaging probes or delivery vehicles for living cells, where the entering of J-type aggregates could be monitored directly with fluorescence microscopy. Similarly, a CF$_3$-linked BODIPY derivative can form strongly luminescent J-aggregates, which is in contrast with the quenched condensed-phase photophysics that are typical for BODIPY dyes. The J-aggregates show narrow red-shifted absorption and emission bands, minimal Stokes shift, and increased fluorescence rate constants. However, with the preparation of J-aggregates, the biological imaging application should be further developed [66]. In another case, 1,4-dimethoxy-2,5-di [4-(cyano)styryl]benzene (COPV) can form water-miscible J-aggregate nanoassemblies driven by the cooperation between π−π

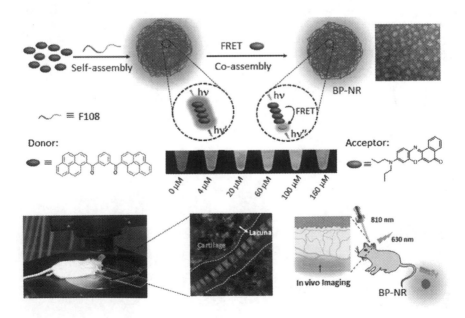

**Fig. 1.8** Schematic illustration and corresponding experimental results of construction of FRET BP1-NR nanosystem and their applications for in vivo imaging

stacking and hydrogen-bonding interactions [67]. The J-aggregates based on COPV were highly emissive in aqueous media, which were also used as efficient two-photon fluorescent contrast agents for bioimaging.

## 1.2.5  In situ Construction

Nature can modulate the biological processes in spatiotemporal way, which involves the direct manipulation of a material, especially a supramolecular material by a stimulus over multiple length and time scales. Inspired by nature, in situ construction/reconstruction of nanomaterials strategy in biological conditions is developed to execute diagnostics and/or therapeutics with signal transmutation to solve unpredictable changes (instability) of the supramolecular nanomaterials under complicated and multivariate physiological environments when applied in vitro and in vivo. Some specific binding interactions, biocompatible reactions and pH, enzyme under specific biological conditions were applied to induce supramolecular interaction changes, such as hydrophobic interactions, resulting in in situ assembly and signal enhancement. The in situ construction was categorized as disassembly, assembly, and reassembly.

### 1.2.5.1  Disassembly

It is well known that fluorescence would be quenched when aggregates for most fluorescence dyes. As a result, the disassembly of fluorescence dyes is a turn-on strategy, which could be utilized for bioimaging and diagnostics (Fig. 1.9). Hamachi developed protein–ligand interactions regulated disassembly to achieve protein recognition with accompanied fluorescence signal changes [68–70]. Tumor-specific or tissue-selective protein biomarkers overexpressed on cell surfaces are important targets for basic biological research and therapeutic applications. The identification of these proteins offers direct applications in environmental monitoring and therapeutics. They carried out specific protein detection by using self-assembled fluorescent nanoprobes consisting of a hydrophilic protein ligand and a hydrophobic BODIPY fluorophore (1–5 ∼ 1–11) [71].

Wang and Chen et al. reported a co-assembled nanovesicle of two BODIPY amphiphiles with distinct emission color. The fluorescence of both BODIPY dyes is completely quenched and recovered upon disassembly driven by solvent variation or cell uptake. Accordingly, the co-assembled nanoaggregates were demonstrated for dual-color imaging of live cells. Furthermore, dual-color and ratiometric fluorescence imaging of live cells indicated that the disaggregated dyes are localized on the endoplasmic reticulum [72]. In the same group, Wang and Qiao et al. demonstrated BP disassembly modulated by a pH-sensitive polymer nanocarrier [73]. The BP was conjugated with hydrophilic poly(amino ester)s (as a pH-sensitive carrier, P-BP). The P-BP could self-assemble into nanoparticles through

| 1- | 1 | 2 | 3 | 4 | 5 | 6 | 7 | 8 | 9 | 10 | 11 |
|---|---|---|---|---|---|---|---|---|---|---|---|
| $R_1$ | 3 | 3 | 4 | 2 | 1 | 1 | 1 | 1 | 1 | 3 | 3 |
| $R_2$ | - | - | - | - | - | - | 1 | 2 | 5 | 3 | 4 |
| $R_3$ | 1 | 3 | 1 | 1 | 1 | 2 | 2 | 2 | 3 | 2 | 2 |
| Linker | 2 | 2 | 3 | 3 | 1 | 3 | 3 | 3 | 3 | 3 | 3 |
| x | 5 | 5 | 10 | 10 | 10 | 10 | 10 | 10 | 10 | 10 | 10 |
| m | - | - | 5 | 5 | 5 | 5 | 5 | 5 | 5 | 5 | 5 |
| n | 1 | 1 | - | - | - | - | - | - | - | - | - |

R₁

1                    2                    3                    4

R₂

1                    2                    3                    4                    5

R₃

1                    3                    4

Linker

1                    2                    3

◀**Fig. 1.9** Protein detection developed by I. Hamachi, the fluorescence probe quenched when self-assembled in aqueous solution. The fluorescence on after disassembled due to the protein recognition

hydrophobic interactions at neutral pH, where the BP aggregated as the hydrophobic core with green fluorescence and polymer as the hydrophilic shell. As the pH decreased, the protonation of polymer chains induced the swelling of nanoparticles with disaggregation of BP. As a result, blue fluorescence of BP monomer increased with decrease of green fluorescence from BP aggregates due to the disassembly. The disassemble nanoparticles could in situ monitor pH values in living cells.

Through protein binding, the supramolecular self-assemblies could be disassembly for sensing, drug delivery, and diagnostics. The structural and functional aspects of assemblies formed from amphiphilic dendrimers can show reversible assembly/disassembly depending on the nature of the solvent medium [74–76]. S. Thayumanavan introduces a new approach of protein-responsive supramolecular disassembly for detection or specific-responsive release of drugs. The amphiphilic dendrimers self-assembled into nanomicelle with hydrophobic guest molecules in their interiors, the stability of which is dependent on the relative compatibility of the hydrophilic and hydrophobic functionalities with water, often referred to as hydrophilic–lipophilic balance (HLB). The protein binds the ligands on amphiphilic dendrimers and induces the changes of HLB, resulting in the disassembly and drug release. In another example, a detection method for specific protein is achieved by using this method. The ligand-tethered fluorophore/quencher pairs are encapsulated in the micellar assemblies through noncovalent interactions, where the fluorescence is in the "off" state. Protein binding-induced dissociation of the ligand-fluorophore combination away from micelles turns the fluorescence to the "on" state. Based on the fluorescence signals, the protein can be recognized and detected. The detection of protein can provide a unique opportunity to explore imbalances in protein activity, which are the primary reasons for most human pathology [76].

## 1.2.5.2  Assembly

The in situ self-assembly is generally realized from the small molecule with hydrophobic aggregatable units and hydrophilic responsive groups (prevent the aggregation due to hydrophilicity and steric hindrance). The hydrophobic aggregatable units self-assemble when the hydrophilic group left upon physiological condition stimuli. The strategy was proposed by Xu and his coworkers, who are experts on hydrogel and developed supramolecular hydrogelation inside living cells in situ [77]. They prepared a precursor of fluorescent hydrogelator, which consists of a fluorophore, 4-nitro-2,1,3-benzoxadiazole (NBD), and a phosphorous ester on the tyrosine residue of a small peptide. The NBD group is a fluorophore known to give more intense fluorescence in hydrophobic environment than in water. The alkaline phosphatase (ALP) catalyzed dephosphorylation can afford a more hydrophobic hydrogelator than the precursor, which is able to self-assemble in water to form nanofibers and to result in the hydrogel. Therefore, the individual hydrogelators exhibit low fluorescence, and the nanofibers of the hydrogelators display bright fluorescence. On the one hand, the strategy allowed the evaluation of intracellular self-assembly, the dynamics, and the localization of the nanofibers of the hydrogelators in live cells. On the other hand, the in situ self-assembly with enhanced fluorescence showed its potential for wash-free cell imaging method. This approach explores supramolecular chemistry inside cells and may lead to new insights, processes, or materials at the interface of chemistry and biology [78].

Rao and Liang reported a condensation reaction-induced hydrophobicity increase, resulting in aggregation to form nanostructures. The condensation reaction can be controlled by reduction or enzyme under physiological conditions with fluorescence readout. The typical condensation reaction could be accomplished between 1,2-aminothiol and 2-cyanobenzothiazole in vitro and in living cells under the control of either pH, disulfide reduction, or enzymatic cleavage [79–82]. The self-assembled nanostructures of the condensation products were fully character-ized, which could be controlled by tuning the structure of the monomers. By applying this condensation reaction-induced assembly, they monitored the caspase-3/7 activity in vitro and in vivo with fluorescence imaging technique [81]. The fluorescent small molecule Cy5.5 was introduced into condensation reaction system, and self-assembled with the condensation reaction to with the enhanced fluorescence. The caspase-3/7 activities were imaged successfully in both apoptotic cells and tumor tissues using three-dimensional structured illumination microscopy.

Tang et al. developed the AIE NPs for fluorescence bioimaging; furthermore, they constructed AIE-based supramolecular aggregates in living cells, where the AIE molecules were released from their conjunction with peptide. They combined DEVD peptide and TPE modules to image the caspase-3 activity [83]. Later on, they extended the AIE detection system by combining the chemotherapeutic drug and biological evaluated chemotherapeutic effect [84]. They synthesized a func-tionalized Pt(IV) prodrug with a cyclic arginine–glycine–aspartic acid (cRGD) tripeptide for targeting integrin $\alpha_v\beta_3$ overexpressed cancer cells and an apoptosis sensor which is composed of TPS fluorophore with AIE characteristics and a caspase-3 enzyme specific Asp-Glu-Val-Asp (DEVD) peptide. The Pt(IV) prodrug can be selectively and effectively uptaken by $\alpha_v\beta_3$ integrin overexpressed cancer cells. Then, the Pt(IV) prodrug can be reduced to produce active Pt(II) drug and release the apoptosis sensor TPS-DEVD simultaneously in cells. The cell apoptosis can be induced by Pt(II) drug and detected by DEVD-TPS, which is cleaved by caspase-3 enzyme and aggregated with fluorescence enhancement. Such noninva-sive and real-time imaging of drug-induced apoptosis in situ can be used as an indicator for early evaluation of the therapeutic responses of a specific anticancer drug. The other enzymes such as cathepsin B [85, 86] and alkaline phosphatase [87] are also used to trigger the aggregation of AIE molecule for in situ imaging.

In the Wang group, the bis-pyrene was developed for J-type aggregate with strong emission in water. They reported supramolecular approach to realize the in situ formation of nanoassemblies in living cells, which involved the pH-responsive amphiphilic carrier and aggregatable monomer BP [88]. The nanocarrier delivered the BP to the lysosomes of cells. In the acidic lysosomes, the bis-pyrene monomers were released and self-aggregated with turn-on fluorescence. The bis(pyrene) derivatives (BP, aggregatable monomer) were molecularly dispersed in the hydrophobic domains of pH-responsive amphiphilic poly(β-amino esters) (PbAE) micelles (responsive carrier) at pH 7.4 via hydrophobic interactions. BP was released and self-aggregated with significant fluorescence enhancement due to the formation of aggregates at pH 5.5 in phosphate buffered solution or serum free cell culture medium (opti-MEM). The release and aggregation of BPs with 6-fold fluorescence enhancement were accomplished within 15 s at 15 μM. Based on unique properties of the supramolecular system, the in situ self-aggregation process of BPs with turn-on fluorescence was successfully achieved in living cells. First, BPs dispersed micelles were accumulated in the lysosomes. Then, the BPs were released and self-aggregated with turn-on fluorescence upon the formation of nanoaggregates in acidic lysosomes. They have demonstrated a supramolecular system to achieve in situ self-aggregation in living cells. The BPs could be transferred into the lysosomes of living cells by dispersing into pH-responsive PbAE micelles via endocytosis. In acidic environment of the lysosomes, the BPs were released and self-aggregated in situ, which was validated by turn-on fluorescence and TEM images of BPs nanoaggregates inside Hela cells. This designed supramolecular strategy is a versatile platform for incorporation of aggregatable monomers into cellular environment-responsive carrier modules, leading to in situ self-aggregation in living objects, which will open a new avenue for highly sensitive molecular diagnostics in the future.

In the same group, they utilized an enzymatic method to modulate the BP aggregates in the living cells to achieve the real-time monitoring autophagy [89]. The intelligent building blocks (DPBP) were composed by a bulky dendrimer carrier, a bis(pyrene) derivatives (BP) as a signal molecule, and a peptide as a responsive unit that can be cleaved by an autophagy-specific enzyme, i.e., ATG4. DPBP maintains the quenched fluorescence in monomer state. However, the responsive peptide is specifically tailored upon activation of autophagy, resulting in self-aggregation of BP residues and 30-fold enhanced fluorescence. By measuring the fluorescent signal values, the autophagic level could be quantitatively evaluated.

### 1.2.5.3 Reassembly

Molecular self-assembly plays an important role in construction of a variety of functional high-ordered nanostructures with precise arrangements in biological systems [90, 91]. The well-defined structure and morphology impart the feature functionality to self-assemblies. Interestingly, smart self-assemblies formed by peptide-based building blocks showed morphology transformation upon stimuli such as enzyme, [92, 93] pH, [94] photoirradiation [95, 96]. The advantages of each morphology can be realized by in situ transformation in vivo for high-performance biological applications. For examples, drug release during the transformation with good controllability in certain time can enhance pharmacology and drug efficacy; [94, 97] the nanoparticles would show prolonged retention effect when they were precisely controlled in transformation into nanofibers [93].

For instance, Giannechi reported the enzyme-responsive spherical nanoparticles form an aggregate-like scaffold or micrometer scale assemblies when activated by metalloproteinase. They utilized enzyme-responsive material to present an in situ reassembly of nanoparticles [92]. First, they described an enzyme-programmed tissue-targeted nanoparticle probe and utilized a Förster resonance energy transfer (FRET)-based assay for monitoring particle accumulation. Generation of a FRET signal provided evidence that the nanoparticles had undergone an enzyme directed aggregation process in tumor tissue generating a slow clearing, self-assembled "implant" of polymeric material within the tissue. Later on, they have utilized Alexa Fluor 647-conjugated peptide polymeric nanoparticles as probes for whole mouse imaging and show extended tumor retention via morphological aggregation in response to MMP enzyme cleavage. Furthermore, they provided compelling evidence that this accumulation process is due to assembly of nanometer particles into larger-scale aggregates by employing super-resolution fluorescence microscopy (stochastic optical reconstruction microscopy, STORM) to study tumor tissue slices ex vivo. They observed fluorescent aggregates in targeted tumor tissues within 1 h that were retained for at least 1 week via detailed tissue-slice analysis coupled with whole animal NIR-fluorescence imaging. Most importantly, particles designed to resist reaction with MMPs were cleared from tumor tissues within 1 h as observed in both in vivo and ex vivo STORM and confocal fluorescence analysis of tissue slices. They validated that the materials had passively diffused into the tumors following injection and then undergone further aggregation with a size increase, which trapped the material within the extracellular space within the tissue.

At the same time, they utilized the same strategy to target the myocardial infarction (MI) as well for achieving prolonged retention of a material in an acute MI via intravenous injection [93]. The nanoparticles were designed to respond to enzymatic stimuli matrix metalloproteinases (MMPs) present in the acute MI resulting in a morphological transition from discrete micellar nanoparticles into network-like scaffolds. The fluorescent nanoparticle is composed of brush peptide–polymer amphiphiles (PPAs) based on a polynorbornene backbone with peptide sequences specific for recognition of MMP-2 and MMP-9. IV delivery allows the enzyme-responsive nanoparticles to freely circulate in the bloodstream until reaching the infarct through the leaky post-MI vasculature, where they assemble and remain within the injured site for up to 28 d post-injection. For the MI application described here, the polymer design was modified to optimize net polymer amphiphilicity. They reduced the degree of polymerization of the hydrophilic block and degree of conjugation of hydrophilic peptide such that the hydrophilic weight fraction was 0.45 instead of 0.55, which increased the responsiveness of the system by decreasing the peptide brush density within formulated micelles of the same size. The in situ assembly in targeted tissue provided a promising approach for long-term imaging and sustained delivery of therapeutics.

Stupp's group prepared pH-responsive peptide amphiphiles, which self-assembled into nanofibers at pH of 7.5 and transformed into nanoparticles with the decreasing pH [94]. By incorporating an oligo-histidine $H_6$ sequence with aliphatic tail on the N-terminus or near the C-terminus, they developed two peptide amphiphiles that self-assembled into distinct morphologies on the nanoscale, either as nanofibers or spherical micelles. Upon protonation of the histidine residues in acidic solutions (pH 6.0 and 6.5), the nanofibers of peptide amphiphiles could transform into nanoparticles at pH 6.0. Meanwhile, the sphere-forming peptide amphiphiles disassembled with impaired hydrogen-bonds. Interestingly, the $H_6$-based nanofiber assemblies encapsulated camptothecin (CPT) with up to 60% efficiency, a sevenfold increase in CPT encapsulation relative to spherical micelles. The peptide amphiphiles were labeled with AF680 to image the biodistribution to understand the effect of shape and pH-dependent assemblies on tumor accumulation in vivo. The results indicated the high accumulation of peptide amphiphiles in mammary tumor as determined by tumor fluorescence. Additionally, pH-sensitive nanofibers showed improved tumor accumulation over both spherical micelles and nanofibers that did not change morphologies in acidic environments. The study demonstrated that the morphological transitions upon changes in pH of supramolecular nanostructures affect drug encapsulation and tumor accumulation.

Wang et al. reported a glutathione (GSH)-based bio-orthogonal reaction that triggers reconstruction of self-assembled nanostructures, resulting in turn-on of dual-color fluorescence signals in living cells [98]. They prepared a cyanine-based amphiphilic dye (BP-S-Cy) with a bis(pyrene) substitution. The BP-S-Cy was initially self-assembled into nanoparticles with quenched fluorescence in water. Second, upon BP-S-Cy was selectively cleaved and substituted by GSH, the released hydrophobic BP unit self assembled into larger nanoaggregates with almost 36-fold fluorescence enhancement with emission maxima at 520 nm. Meanwhile, the newly formed GSH labeled Cy (Cy-GSH) exhibited 30-fold fluorescence enhancement at NIR region ($\lambda_{em}$ 820 nm). Finally, this new GSH triggered bio-orthogonal reaction and resultant reconstruction of self-assembled nanostructures were validated in living cells. This smart nanoprobe was utilized to image the GSH in living cells with dual-wavelength emission.

The understanding of formation and structure-evolutionary mechanism of self-assembled superstructures is still a challenge. Wang and his coworkers designed and prepared a series of supramolecular building blocks (BP-KLVFFG-PEG, BKP) composed by an aromatic bis-pyrene motif, a hydrogen-bonding motif, and a hydrophilic poly ethylene glycol (PEG) chain and investigated their self-assembly processes systematically (Fig. 1.10) [99]. BP was chosen as hydrophobic core to induce aggregation and monitor the aggregation by fluorescence signals from BP upon aggregation. The peptide with a sequence of Lys-Leu-Val-Phe-Phe (KLVFF) originated from the segments of amyloid β has been proven to be a highly fibrillation by H-bonds [100]. The polymeric peptide (KLVFF) was chosen as peptide scaffolds for inducing fiber formation. It is well known that hydrophilic–lipophilic balance (HLB) affects the self-assembly process, and resultant structures and morphologies of supramolecular nanomaterials [101].

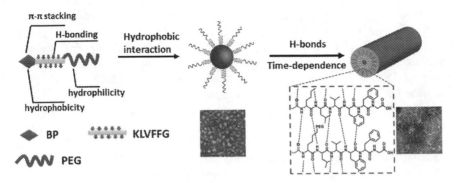

**Fig. 1.10** Schematic illustration of In situ formation of nanofibers from nanoparticles of BP-KLVFFG-PEG polymer peptide and corresponding TEM images

The PEG with different molecular weights (Mw = 368, 1000, 2000) was chosen as hydrophilic chains to modulate HLB of the target molecules. The study revealed that the hydrophobic and $\pi-\pi$ interactions initially drove the self-assembly of BKP into nanoparticles with $J$-type fashion, and the H-bonding interactions further drove in situ spontaneous morphology transformation into nanofibers. The longer was hydrophilic chain, the faster transformation was from nanoparticles to nanofibers. The elucidation of BKP assemblies from the molecular level to supramolecular nanosystems constitutes a bridge between molecular structures and their morphologies.

Based on the above systematical investigation of morphology transformation, the metal ions-regulated morphology transformation was proposed and confirmed in the same group [102]. It is well known that RGD has specific binding activity to integrin $\alpha_v\beta_3$, which is mainly driven by the coordination interactions between RGD and the metal ions ($Ca^{2+}$, $Mg^{2+}$) at "metal ion-dependent adhesion site" (MIDAS) [103–105]. They designed and prepared modular peptide-based building blocks (BKR) composed by an aromatic bis-pyrene (BP) motif, a Lys-Leu-Val-Phe-Phe (KLVFF) motif, and a divalent cation ($Ca^{2+}$) binding motif (Arg-Gly-Asp, RGD). The study demonstrated metal ion-mediated structural evolution of peptide-based superstructures on specific cancer cell surfaces for interfering cellular viability by applying the RGD and integrin $\alpha_v\beta_3$ interactions. Actually, the BKR could light up calcium alginate beads and U87 cells through binding with the $Ca^{2+}$ on their surfaces, respectively. More importantly, BKR on the U87 cell surfaces formed nanofibers in situ from nanoparticles and killed the cells. The bio-inspired morphology transformation of BKR peptide on the cell surfaces displays a great potential for cancer theranostics.

## 1.3 Summary and Outlook

In this review, recent progress of supramolecular self-assembled nanomaterials for fluorescence bioimaging was summarized. Supramolecular nanoassemblies as contrast agents have many advantages compared to small molecules, such as high stability, high sensitivity, and specificity. However, the preparation reliability of self-assembled nanomaterials is a key issue, which is due to the difficulty in controlling the supramolecular interactions. It is possible to transform the morphologies, structures, and corresponding fluorescence properties when encountering in biointerfaces with complicated physiological condition in vivo, ascribed to the dynamic nature of self-assembled nanomaterials.

For ex situ self-assembled nanomaterials, the dye-dispersed polymeric or semiconducting polymeric nanoparticles have been proven to be simple and effective supramolecular approach to construct nanoprobes/contrast agents for general fluorescence materials. The well-controlled nanomaterials with definite parameters such as size, shape, charge, payload, and surface chemistry are necessary for reliable application. The aggregation of fluorescence units should be precisely tuned to keep high fluorescence quantum yields by incorporating bulky groups, introducing moieties to induce intramolecular and intermolecular energy transfer. The AIE and J-type aggregation nanoparticles take advantage of fluorescence increase nature during aggregation process. For all the ex situ self-assembled nanomaterials, the stability is highly disable to face the complicated physiological conditions. They could meet the requirement of high-performance fluorescence contrast agents for clinical application.

On the other hand, the in situ self-assembly is much more interesting field that draws great attention from fundamental research. The in situ self-assembly can be intelligent to monitor the biological process, activity of biomolecules, and disease diagnosis in vivo. More importantly, in situ self-assembly is generally achieved under natural stimuli, such as enzyme, pH, and redox in specific physiological and pathological region. It will help the construction of nanoprobes or release of drugs effective in therapeutics and diagnostics due to the selective and efficient homing of material to specific tissues and biomolecular targets. For example, the in situ assembly showed long retention time, which was called aggregation-induced retention (AIR) effect, resulting in high performance of imaging and treatment. There is an extraordinary need of powerful new tools and strategies for characterizing the dynamics, morphology, and behavior of in situ self-assembled nanomaterials, which can identify their specific and useful physical properties from normal tissue. Furthermore, the design of sophisticated precursor for self-assembly would be a large challenge for the in situ assembly with fluorescence signal difference.

For all the self-assembled nanomaterials, further development bioimaging applications will be mainly focused on the FR/NIR emitters with narrow band emission. Various signal amplification strategies, such as FRET and metal-enhanced fluorescence, could be used to further improve their brightness for fluorescence imaging. The biodistribution, such as liver and spleen, and their toxicity in vivo following systemic administration of self-assembled nanomaterials need to be evaluated through standardized, validated methods. Furthermore, the morphology and structure transformation followed by the disassembly and excretion of self-assembled nanomaterials in vivo should be paid great effort. The self-assembled nanomaterials with disassembility and degradability with high biocompatibility and low toxicity by systematically evaluation could be applied to clinical diagnostics and therapeutics of diseases. By collaboration with the multidisciplinary researchers from chemistry, biology, and medicine, we believe that the self-assembled nanomaterials will lead to benefits to all of us in the future.

# References

1. Lehn J-M (1988) Supramolecular chemistry-scope and perspectives molecules, supermolecules, and molecular devices (Nobel Lecture). Angew Chem Int Ed 27(1):89–112
2. Li F, Lu J, Kong X, Hyeon T, Ling D (2017) Dynamic nanoparticle assemblies for biomedical applications. Adv Mater:1605897
3. Wang L, L-l Li, Y-s Fan, Wang H (2013) Host–guest supramolecular nanosystems for cancer diagnostics and therapeutics. Adv Mater 25(28):3888–3898
4. de Silva AP, Gunaratne HQN, Gunnlaugsson T, Huxley AJM, McCoy CP, Rademacher JT, Rice TE (1997) Signaling recognition events with fluorescent sensors and switches. Chem Rev 97(5):1515–1566
5. Michalet X, Pinaud FF, Bentolila LA, Tsay JM, Doose S, Li JJ, Sundaresan G, Wu AM, Gambhir SS, Weiss S (2005) Quantum dots for live cells, in vivo imaging, and diagnostics. Science 307(5709):538–544
6. de Chermont QlM, Chaneac C, Seguin J, Pelle F, Maitrejean S, Jolivet J-P, Gourier D, Bessodes M, Scherman D (2007) Nanoprobes with near-infrared persistent luminescence for in vivo imaging. Proc Natl Acad Sci USA 104(22):9266–9271
7. He S, Song B, Li D, Zhu C, Qi W, Wen Y, Wang L, Song S, Fang H, Fan C (2010) A graphene nanoprobe for rapid, sensitive, and multicolor fluorescent DNA analysis. Adv Funct Mater 20(3):453–459
8. Ding D, Li K, Liu B, Tang BZ (2013) Bioprobes based on AIE Fluorogens. Acc Chem Res 46(11):2441–2453
9. Bao B, Fan Q, Wang L, Huang W (2014) Recent advances in fluorescent conjugated polymer nanoparticles. Sci China Chem 44(3):282–294
10. Chen M, Yin M (2014) Design and development of fluorescent nanostructures for bioimaging. Prog Polym Sci 39(2):365–395
11. Razgulin A, Ma N, Rao J (2011) Strategies for in vivo imaging of enzyme activity: an overview and recent advances. Chem Soc Rev 40(7):4186–4216
12. Tao Z, Hong G, Shinji C, Chen C, Diao S, Antaris AL, Zhang B, Zou Y, Dai H (2013) Biological imaging using nanoparticles of small organic molecules with fluorescence emission at wavelengths longer than 1000 nm. Angew Chem Int Ed 52(49):13002–13006
13. Wu C, Schneider T, Zeigler M, Yu J, Schiro PG, Burnham DR, McNeill JD, Chiu DT (2010) Bioconjugation of ultrabright semiconducting polymer dots for specific cellular targeting. J Am Chem Soc 132(43):15410–15417
14. Peng HS, Stolwijk JA, Sun LN, Wegener J, Wolfbeis OS (2010) A nanogel for ratiometric fluorescent sensing of intracellular pH values. Angew Chem Int Ed 49(25):4246–4249
15. Wei H, Zhuo RX, Zhang XZ (2013) Design and development of polymeric micelles with cleavable links for intracellular drug delivery. Prog Polym Sci 38(3–4):503–535

16. Yin MZ, Shen J, Pisula W, Liang MH, Zhi LJ, Müllen K (2009) Functionalization of self-assembled hexa-peri-hexabenzocoronene fibers with peptides for bioprobing. J Am Chem Soc 131(41):14618–14619

17. Luo J, Lei T, Wang L, Ma YG, Cao Y, Wang J, Pei J (2009) Highly fluorescent rigid supramolecular polymeric nanowires constructed through multiple hydrogen bonds. J Am Chem Soc 131(6):2076–2077

18. Yin MZ, Feng CL, Shen J, Yu YM, Xu ZJ, Yang WT, Knoll W, Müllen K (2011) Dual-responsive interaction to detect DNA on template-based fluorescent nanotubes. Small 7 (12):1629–1634

19. Zhang D, Zhao Y-X, Qiao Z-Y, Mayerhöffer U, Spenst P, Li X-J, Würthner F, Wang H (2014) Nano-confined squaraine dye assemblies: new photoacoustic and near-infrared fluorescence dual-modular imaging probes in vivo. Bioconj Chem 25(11):2021–2029

20. Bunschoten A, Buckle T, Kuil J, Luker GD, Luker KE, Nieweg OE, van Leeuwen FWB (2012) Targeted non-covalent self-assembled nanoparticles based on human serum albumin. Biomaterials 33(3):867–875

21. Chen Q, Liang C, Wang X, He J, Li Y, Liu Z (2014) An albumin-based theranostic nano-agent for dual-modal imaging guided photothermal therapy to inhibit lymphatic metastasis of cancer post surgery. Biomaterials 35(34):9355–9362

22. Chen Q, Wang C, Zhan Z, He W, Cheng Z, Li Y, Liu Z (2014) Near-infrared dye bound albumin with separated imaging and therapy wavelength channels for imaging-guided phototothermal therapy. Biomaterials 35(28):8206–8214

23. Sheng Z, Hu D, Zheng M, Zhao P, Liu H, Gao D, Gong P, Gao G, Zhang P, Ma Y, Cai L (2014) Smart human serum albumin-indocyanine green nanoparticles generated by programmed assembly for dual-modal imaging-guided cancer synergistic phototherapy. ACS Nano 8(12):12310–12322

24. Gao F-P, Lin Y-X, Li L-L, Liu Y, Mayerhoeffer U, Spenst P, Su J-G, Li J Y, Würthner F, Wang H (2014) Supramolecular adducts of squaraine and protein for noninvasive tumor imaging and photothermal therapy in vivo. Biomaterials 35(3):1004–1014

25. Wu C, Hansen SJ, Hou Q, Yu J, Zeigler M, Jin Y, Burnham DR, McNeill JD, Olson JM, Chiu DT (2011) Design of highly emissive polymer dot bioconjugates for in vivo tumor targeting. Angew Chem Int Ed 50(15):3430–3434

26. Ye F, Wu C, Jin Y, Chan Y-H, Zhang X, Chiu DT (2011) Ratiometric temperature sensing with semiconducting polymer dots. J Am Chem Soc 133(21):8146–8149

27. Chan Y-H, Ye F, Gallina ME, Zhang X, Jin Y, Wu IC, Chiu DT (2012) Hybrid semiconducting polymer dot-quantum dot with narrow-band emission, near infrared fluorescence, and high brightness. J Am Chem Soc 134(17):7309–7312

28. Yu J, Wu C, Zhang X, Ye F, Gallina ME, Rong Y, Wu IC, Sun W, Chan Y-H, Chiu DT (2012) Stable functionalization of small semiconducting polymer dots via covalent cross-linking and their application for specific cellular imaging. Adv Mater 24(26):3498–3504

29. Peng H-S, Chiu DT (2015) Soft fluorescent nanomaterials for biological and biomedical imaging. Chem Soc Rev 44(14):4699–4722

30. Xiong L, Shuhendler AJ, Rao J (2012) Self-luminescing BRET-FRET near-infrared dots for in vivo lymph-node mapping and tumour imaging. Nat Commun 3:1193

31. Shuhendler AJ, Pu K, Cui L, Uetrecht JP, Rao J (2014) Real-time imaging of oxidative and nitrosative stress in the liver of live animals for drug-toxicity testing. Nat Biotechnol 32 (4):373–380

32. Palner M, Pu K, Shao S, Rao J (2015) Semiconducting polymer nanoparticles with persistent near-infrared luminescence for in vivo optical imaging. Angew Chem Int Ed 54(39):11477–11480

33. Feng L, Zhu C, Yuan H, Liu L, Lv F, Wang S (2013) Conjugated polymer nanoparticles: preparation, properties, functionalization and biological applications. Chem Soc Rev 42 (16):6620–6633

34. Jeong K, Park S, Lee YD, Lim CK, Kim J, Chung BH, Kwon IC, Park CR, Kim S (2013) Conjugated polymer/photochromophore binary nanococktails: bistable photoswitching of near-infrared fluorescence for in vivo imaging. Adv Mater 25(39):5574–5580
35. Jin Y, Ye F, Zeigler M, Wu C, Chiu DT (2011) Near-infrared fluorescent dye-doped semiconducting polymer dots. ACS Nano 5(2):1468–1475
36. Danhier F, Ansorena E, Silva JM, Coco R, Le Breton A, Preat V (2012) PLGA-based nanoparticles: an overview of biomedical applications. J Control Release 161(2):505–522
37. Li K, Pan J, Feng SS, Wu AW, Pu KY, Liu YT, Liu B (2009) Generic strategy of preparing fluorescent conjugated-polymer-loaded poly(DL-lactide-co-glycolide) nanoparticles for targeted cell imaging. Adv Funct Mater 19(22):3535–3542
38. Li K, Liu YT, Pu KY, Feng SS, Zhan RY, Liu B (2011) Polyhedral oligomeric silsesquioxanes-containing conjugated polymer loaded PLGA nanoparticles with trastuzumab (herceptin) functionalization for HER2-positive cancer cell detection. Adv Funct Mater 21(2):287–294
39. Hong Y, Lam JWY, Tang BZ (2011) Aggregation-induced emission. Chem Soc Rev 40 (11):5361–5388
40. Qin A, Lam JWY, Tang BZ (2012) Luminogenic polymers with aggregation-induced emission characteristics. Prog Polym Sci 37(1):182–209
41. Hu R, Leung NLC, Tang BZ (2014) AIE macromolecules: syntheses, structures and functionalities. Chem Soc Rev 43(13):4494–4562
42. Mei J, Hong Y, Lam JWY, Qin A, Tang Y, Tang BZ (2014) Aggregation-induced emission: the whole is more brilliant than the parts. Adv Mater 26(31):5429–5479
43. Kwok RTK, Leung CWT, Lam JWY, Tang BZ (2015) Biosensing by luminogens with aggregation-induced emission characteristics. Chem Soc Rev 44(13):4228–4238
44. Liang J, Tang BZ, Liu B (2015) Specific light-up bioprobes based on AIEgen conjugates. Chem Soc Rev 44(10):2798–2811
45. Mei J, Leung NLC, Kwok RTK, Lam JWY, Tang BZ (2015) Aggregation-induced emission: together we shine, united we soar! Chem Rev 115(21):11718–11940
46. Wang H, Zhao E, Lam JWY, Tang BZ (2015) AIE luminogens: emission brightened by aggregation. Mater Today 18(7):365–377
47. Zhao Z, He B, Tang BZ (2015) Aggregation-induced emission of siloles. Chem Sci 6 (10):5347–5365
48. Yu Y, Feng C, Hong Y, Liu J, Chen S, Ng KM, Luo KQ, Tang BZ (2011) Cytophilic fluorescent bioprobes for long-term cell tracking. Adv Mater 23(29):3298–3302
49. Ma H, Qi C, Cheng C, Yang Z, Cao H, Yang Z, Tong J, Yao X, Lei Z (2016) AIE-active tetraphenylethylene cross-linked N-isopropylacrylamide polymer: a long-term fluorescent cellular tracker. ACS Appl Mater Interfaces 8(13):8341–8348
50. Wang Z, Chen S, Lam JWY, Qin W, Kwok RTK, Xie N, Hu Q, Tang BZ (2013) Long-term fluorescent cellular tracing by the aggregates of AIE bioconjugates. J Am Chem Soc 135 (22):8238–8245
51. Qin W, Ding D, Liu J, Yuan WZ, Hu Y, Liu B, Tang BZ (2012) Biocompatible nanoparticles with aggregation-induced emission characteristics as far-red/near-infrared fluorescent bioprobes for in vitro and in vivo imaging applications. Adv Funct Mater 22 (4):771–779
52. Ding D, Li K, Qin W, Zhan R, Hu Y, Liu J, Tang BZ, Liu B (2013) Conjugated polymer amplified far-red/near-infrared fluorescence from nanoparticles with aggregation-induced emission characteristics for targeted in vivo imaging. Adv Healthcare Mater 2(3):500–507
53. Li K, Jiang Y, Ding D, Zhang X, Liu Y, Hua J, Feng S-S, Liu B (2011) Folic acid-functionalized two-photon absorbing nanoparticles for targeted MCF-7 cancer cell imaging. Chem Commun 47(26):7323–7325
54. Li K, Qin W, Ding D, Tomczak N, Geng J, Liu R, Liu J, Zhang X, Liu H, Liu B, Tang BZ (2013) Photostable fluorescent organic dots with aggregation-induced emission (AIE dots) for noninvasive long-term cell tracing. Scientif Rep 3:1150

55. Qin W, Li K, Feng G, Li M, Yang Z, Liu B, Tang BZ (2014) Bright and photostable organic fluorescent dots with aggregation-induced emission characteristics for noninvasive long-term cell imaging. Adv Funct Mater 24(5):635–643

56. Gao Y, Feng G, Jiang T, Goh C, Ng L, Liu B, Li B, Yang L, Hua J, Tian H (2015) Biocompatible nanoparticles based on diketo-pyrrolo-pyrrole (DPP) with aggregation-induced red/NIR emission for in vivo two-photon fluorescence imaging. Adv Funct Mater 25(19):2857–2866

57. Shao A, Xie Y, Zhu S, Guo Z, Zhu S, Guo J, Shi P, James TD, Tian H, Zhu W-H (2015) Far-red and near-IR AIE-active fluorescent organic nanoprobes with enhanced tumor-targeting efficacy: shape-specific effects. Angew Chem Int Ed 54(25):7275–7280

58. Geng J, Goh CC, Qin W, Liu R, Tomczak N, Ng LG, Tangde BZ, Liu B (2015) Silica shelled and block copolymer encapsulated red-emissive AIE nanoparticles with 50% quantum yield for two-photon excited vascular imaging. Chem Commun 51(69):13416–13419

59. Geng J, Li K, Ding D, Zhang X, Qin W, Liu J, Tang BZ, Liu B (2012) Lipid-PEG-folate encapsulated nanoparticles with aggregation induced emission characteristics: cellular uptake mechanism and two-photon fluorescence imaging. Small 8(23):3655–3663

60. Ding D, Goh CC, Feng G, Zhao Z, Liu J, Liu R, Tomczak N, Geng J, Tang BZ, Ng LG, Liu B (2013) Ultrabright organic dots with aggregation-induced emission characteristics for real-time two-photon intravital vasculature imaging. Adv Mater 25(42):6083–6088

61. Kaiser TE, Wang H, Stepanenko V, Würthner F (2007) Supramolecular construction of fluorescent J-aggregates based on hydrogen-bonded perylene dyes. Angew Chem Int Ed 46(29):5541–5544

62. Wang H, Kaiser TE, Uemura S, Würthner F (2008) Perylene bisimide J-aggregates with absorption maxima in the NIR. Chem Commun 10:1181–1183

63. Wang L, Li W, Lu J, Zhao Y X, Fan G, Zhang J-P, Wang H (2013) Supramolecular nano-aggregates based on bis(pyrene) derivatives for lysosome-targeted cell imaging. J Phys Chem C 117(50):26811–26820

64. Yang P-P, Yang Y, Gao Y-J, Wang Y, Zhang J-C, Lin Y-X, Dai L, Li J, Wang L, Wang H (2015) Unprecedentedly high tissue penetration capability of co-assembled nanosystems for two-photon fluorescence imaging in vivo. Adv Optical Mater 3(5):646–651

65. Olivier J-H, Widmaier J, Ziessel R (2011) Near-infrared fluorescent nanoparticles formed by self-assembly of lipidic (bodipy) dyes. Chem Eur J 17(42):11709–11714

66. Choi S, Bouffard J, Kim Y (2014) Aggregation-induced emission enhancement of a meso-trifluoromethyl BODIPY via J-aggregation. Chem Sci 5(2):751–755

67. Xu Z, Liao Q, Wu Y, Ren W, Li W, Liu L, Wang S, Gu Z, Zhang H, Fu H (2012) Water-miscible organic J-aggregate nanoparticles as efficient two-photon fluorescent nano probes for bio-imaging. J Mater Chem 22(34):17737–17743

68. Hayashi T, Hamachi I (2012) Traceless affinity labeling of endogenous proteins for functional analysis in living cells. Acc Chem Res 45(9):1460–1469

69. Kubota R, Hamachi I (2015) Protein recognition using synthetic small-molecular binders toward optical protein sensing in vitro and in live cells. Chem Soc Rev 44(13):4454–4471

70. Takaoka Y, Ojida A, Hamachi I (2013) Protein organic chemistry and applications for labeling and engineering in live-cell systems. Angew Chem Int Ed 52(15):4088–4106

71. Mizusawa K, Takaoka Y, Hamachi I (2012) Specific cell surface protein imaging by extended self-assembling fluorescent turn-on nanoprobes. J Am Chem Soc 134(32):13386–13395

72. Fan G, Lin Y-X, Yang L, Gao F-P, Zhao Y-X, Qiao Z-Y, Zhao Q, Fan Y-S, Chen Z, Wang H (2015) Co-self-assembled nanoaggregates of BODIPY amphiphiles for dual colour imaging of live cells. Chem Commun 51(62):12447–12450

73. Qiao Z-Y, Hou C-Y, Zhao W-J, Zhang D, Yang P-P, Wang L, Wang H (2015) Synthesis of self-reporting polymeric nanoparticles for in situ monitoring of endocytic microenvironmental pH. Chem Commun 51(63):12609–12612

74. Raghupathi KR, Guo J, Munkhbat O, Rangadurai P, Thayumanavan S (2014) Supramolecular disassembly of facially amphiphilic dendrimer assemblies in response to physical, chemical, and biological stimuli. Acc Chem Res 47(7):2200–2211
75. Amado Torres D, Garzoni M, Subrahmanyam AV, Pavan GM, Thayumanavan S (2014) Protein-triggered supramolecular disassembly: insights based on variations in ligand location in amphiphilic dendrons. J Am Chem Soc 136(14):5385–5399
76. Wang H, Zhuang J, Raghupathi KR, Thayumanavan S (2015) A supramolecular dissociation strategy for protein sensing. Chem Commun 51(97):17265–17268
77. Gao Y, Shi J, Yuan D, Xu B (2012) Imaging enzyme-triggered self-assembly of small molecules inside live cells. Nat Commun 3:1033
78. Gao Y, Berciu C, Kuang Y, Shi J, Nicastro D, Xu B (2013) Probing nanoscale self-assembly of nonfluorescent small molecules inside live mammalian cells. ACS Nano 7(10):9055–9063
79. Dragulescu-Andrasi A, Kothapalli S-R, Tikhomirov GA, Rao J, Gambhir SS (2013) Activatable oligomerizable imaging agents for photoacoustic imaging of furin-like activity in living subjects. J Am Chem Soc 135(30):11015–11022
80. Liang G, Ren H, Rao J (2010) A biocompatible condensation reaction for controlled assembly of nanostructures in living cells. Nat Chem 2(1):54–60
81. Ye D, Shuhendler AJ, Cui L, Tong L, Tee SS, Tikhomirov G, Felsher DW, Rao J (2014) Bioorthogonal cyclization-mediated in situ self-assembly of small-moleculeprobes for imaging caspase activity in vivo. Nat Chem 6(6):519–526
82. Ye D, Shuhendler AJ, Pandit P, Brewer KD, Tee SS, Cui L, Tikhomirov G, Rutt B, Rao J (2014) Caspase-responsive smart gadolinium-based contrast agent for magnetic resonance imaging of drug-induced apoptosis. Chem Sci 5(10):3845–3852
83. Shi H, Kwok RTK, Liu J, Xing B, Tang BZ, Liu B (2012) Real-time monitoring of cell apoptosis and drug screening using fluorescent light-up probe with aggregation-induced emission characteristics. J Am Chem Soc 134(43):17972–17981
84. Yuan Y, Kwok RTK, Tang BZ, Liu B (2014) Targeted theranostic platinum (IV) prodrug with a built-in aggregation-induced emission light-up apoptosis sensor for noninvasive early evaluation of its therapeutic responses in situ. J Am Chem Soc 136(6):2546–2554
85. Han H, Jin Q, Wang Y, Chen Y, Ji J (2015) The rational design of a gemcitabine prodrug with AIE-based intracellular light-up characteristics for selective suppression of pancreatic cancer cells. Chem Commun 51(98):17435–17438
86. Yuan Y, Zhang C-J, Gao M, Zhang R, Tang BZ, Liu B (2015) Specific light-up bioprobe with aggregation-induced emission and activatable photoactivity for the targeted and image-guided photodynamic ablation of cancer cells. Angew Chem Int Ed 54(6):1780–1786
87. Liang J, Kwok RTK, Shi H, Tang BZ, Liu B (2013) Fluorescent light-up probe with aggregation-induced emission characteristics for alkaline phosphatase sensing and activity study. ACS Appl Mater Interfaces 5(17):8784–8789
88. Duan Z, Gao Y-J, Qiao Z-Y, Qiao S, Wang Y, Hou C, Wang L, Wang H (2015) pH-Sensitive polymer assisted self-aggregation of bis(pyrene) in living cells in situ with turn-on fluorescence. Nanotechnology 26(35)
89. Lin Y-X, Qiao S-L, Wang Y, Zhang R-X, An H-W, Ma Y, Rajapaksha RPYJ, Qiao Z-Y, Wang L, Wang H (2017) An in situ intracellular self-assembly strategy for quantitatively and temporally monitoring autophagy. ACS Nano 11(2):1826–1839
90. Carter NA, Grove TZ (2015) Repeat-proteins films exhibit hierarchical anisotropic mechanical properties. Biomacromol 16(3):706–714
91. Feldkamp U, Niemeyer CM (2006) Rational design of DNA nanoarchitectures. Angew Chem Int Ed 45(12):1856–1876
92. Callmann CE, Barback CV, Thompson MP, Hall DJ, Mattrey RF, Gianneschi NC (2015) Therapeutic enzyme-responsive nanoparticles for targeted delivery and accumulation in tumors. Adv Mater 27(31):4611–4615
93. Nguyen MM, Carlini AS, Chien M-P, Sonnenberg S, Luo C, Braden RL, Osborn KG, Li Y, Gianneschi NC, Christman KL (2015) Enzyme-responsive nanoparticles for targeted

accumulation and prolonged retention in heart tissue after myocardial infarction. Adv Mater 27(37):5547–5552

94. Moyer TJ, Finbloom JA, Chen F, Toft DJ, Cryns VL, Stupp SI (2014) pH and amphiphilic structure direct supramolecular behavior in biofunctional assemblies. J Am Chem Soc 136 (42):14746–14752

95. Matson JB, Navon Y, Bitton R, Stupp SI (2015) Light-controlled hierarchical self-assembly of polyelectrolytes and supramolecular polymers. ACS Macro Lett 4(1):43–47

96. Sun H-L, Chen Y, Zhao J, Liu Y (2015) Photocontrolled reversible conversion of nanotube and nanoparticle mediated by -cyclodextrin dimers. Angew Chem Int Ed 54(32):9376–9380

97. Shao H, Parquette JR (2009) Controllable peptide-dendron self-assembly: interconversion of nanotubes and fibrillar nanostructures. Angew Chem Int Ed 48(14):2525–2528

98. An H-W, Qiao S-L, Li L-L, Yang C, Lin Y-X, Wang Y, Qao Z-Y, Wang L, Wang H (2016) Bio-orthogonally Deciphered binary nanoemitters for tumor diagnostics. ACS Appl Mater Interfaces 8(30).19202–19207

99. Yang P-P, Zhao X-X, Xu A-P, Wang L, Wang H (2016) Reorganization of self-assembled supramolecular materials controlled by hydrogen bonding and hydrophilic-lipophilic balance. J Mater Chem B 4(15):2662–2668

100. Esler WP, Stimson ER, Ghilardi JR, Lu Y-A, Felix AM, Vinters HV, Mantyh PW, Lee JP, Maggio JE (1996) Point substitution in the central hydrophobic cluster of a human β-Amyloid congener disrupts peptide folding and abolishes plaque competence. Biochemistry 35(44):13914–13921

101. Molla MR, Prasad P, Thayumanavan S (2015) Protein-induced supramolecular disassembly of amphiphilic polypeptide nanoassemblies. J Am Chem Soc 137(23):7286–7289

102. Xu A-P, Yang P-P, Yang C, Gao Y-J, Zhao X-X, Luo Q, Li X-D, Li L-Z, Wang L, Wang H (2016) Bio-inspired metal ions regulate the structure evolution of self-assembled peptide-based nanoparticles. Nanoscale 8(29):14078–14083

103. Yu YP, Wang Q, Liu YC, Xie Y (2014) Molecular basis for the targeted binding of RGD-containing peptide to integrin αvβ3. Biomaterials 35(5):1667–1675

104. Craig D, Gao M, Schulten K, Vogel V (2004) Structural insights into how the MIDAS ion stabilizes integrin binding to an RGD peptide under force. Structure 12(11):2049–2058

105. Puklin-Faucher E, Vogel V (2009) Integrin activation dynamics between the RGD-binding site and the headpiece hinge. J Biol Chem 284(52):36557–36568

# Chapter 2
# The Self-assembly of Cyanine Dyes for Biomedical Application In Vivo

Hong-Wei An and Man-Di Wang

**Abstract** The development of near-infrared (NIR) fluorescence imaging is expected to have significant impact on future personalized oncology owing to the very low tissue autofluorescence and high tissue penetration depth in the NIR spectrum window. Among conventional fluorescent probes, cyanine dyes, one kind of NIR dyes, exhibit unique advantages in molecular tracing both in vitro and in vivo. Cyanine dye possessed desired features of J-aggregates, such as a narrow absorption band that is bathochromically shifted relative to the monomer band, high fluorescence intensity with small Stokes shift, and exceptional exciton transport capability. The strong absorption of light and efficient exciton transport to analyte traps provide these dye aggregates with an ultrahigh sensitivity. Thus, it is not surprising that cationic cyanine dye aggregates were the dyes of choice for the development of novel fluorescent and colorimetric sensing schemes. Moreover, the use of self-assembled biomaterials in bioapplication could also benefit from improved nanostructural precision. It is well knowledged to construct well-ordered nanostructures in living objectives for reflecting certain physiological activity which is one of the ultimate goals for supramolecular chemistry and self-assembled functional materials. We propose a new "in vivo self-assembly" concept which means the construction of well-ordered self-assemblies from exogenous small molecules in bioenvironments (e.g., cells, tissues or living animals). These self-assemblies usually possess multiple nature mimicking biological functions. So we

H.-W. An (✉)
CAS Center for Excellence in Nanoscience, CAS Key Laboratory for Biomedical Effects of Nanomaterials and Nanosafety, National Center for Nanoscience and Technology (NCNST), No. 11 Beiyitiao, Zhongguancun, Beijing, China
e-mail: anhw@nanoctr.cn

H.-W. An · M.-D. Wang
Institute of High Energy Physics, Chinese Academy of Sciences (CAS), Yuquan Road, Beijing, China
e-mail: wangmd@nanoctr.cn

© Springer Nature Singapore Pte Ltd. 2018
H. Wang and L.-L. Li (eds.), *In Vivo Self-Assembly Nanotechnology for Biomedical Applications*, Nanomedicine and Nanotoxicology,
https://doi.org/10.1007/978-981-10-6913-0_2

work on developing new cyanine dye self-assembly with biofunction. And we are trying to the design of controllable cyanine dye building blocks with biofunctions, tuning the balance of attractive and repulsive intermolecular forces, accurately positioning self-assembly, rapid accumulation, and prediction of their self-assembly behavior in living body.

## 2.1   Introduction

In the last couple of decades, the applications of fluorescent probes in biological and biomedical fields display dramatically increasing. The widely acknowledged sensitivity, designed specificity, and multiple choice of candidates for the fluorescence have replaced radioactivity in many bioapplications, such as bioimaging of bioactive molecules/materials [1, 2] or investigating the structure/dynamics of biological macromolecules including nucleic acids [3, 4], proteins [4] [5]. As concern to living subject, the deep penetration of the excitation and emission peaks and high fluorescence quantum yield are two significant parameters for consideration of fluorescence probes. The redshift of the "imaging window" to 650–1450, where the absorption by tissues and blood is at a minimum, defines the fluorophores as near-infrared fluorescent probes. Meanwhile, the favorable strong signal wavelength should lie in 650–1450 [6]. Right now, most organic fluorescent probes with near-infrared (NIR) excitation and emission profiles are divided into four categories, which are cyanines, phthalocyanines/porphyrins/pyrroles [7] squaraines, and BODIPYs [8]. Although these NIR dyes was found and applicated before decades, to date, only indocyanine green was approved by FDA for medical administration in clinical diagnostics. The biocompatibility and fluorescence property concern of cyanine dyes inspire us to modify or derive these molecules improving their characterizations for better use as fluorescent agent or a NIR absorption agent [6, 9].

Among conventional fluorescent probes, cyanine dyes, one kind of NIR dyes, exhibit unique advantages in molecular tracing both in vitro and in vivo [10]. There are two significant considerations giving priority to NIR dyes in biomedical applications. One is lower interference of high tissue autofluorescence from indigenous biomolecules to NIR emission in living subjects. The other is deeper penetration of NIR photons into tissues and less damage to biological samples. Undoubtedly, NIR fluorescent probes will provide us a powerful tool in depth to understand our body at a molecular level. In actual application, designed NIR probes have been successfully employed for cancer or other serious disease imaging or diagnosis, as well as some other clinical practices.

## 2.2 Characterizations of Cyanine Dyes

### 2.2.1 Chemical Structure and Fluorescence Characteristic of Cyanine Dyes

The cyanine dye is originally found probably by Williams in 1856 [11]. Then, Williams first named this brilliant blue-shaded dye as cyanine (cyanos = blue), which is obtained by reacting quinoline with amyl iodide. The common feature of these cyanine dyes is that they consist of two heterocyclic units in the chemical structure, and the two units are connected by an odd number of methine group (CH) n (with n = 1, 3, 5, …). Until 1920, König introduced the term polymethine dyes, which was named after the specific phenomenon that the colors of these dyes were mainly determined by the length of polymethine chain [12, 13]. The reason is that polymethine chains connect predominantly aromatic heterocyclic donor (D) and acceptor (A) groups in cyanine structure [14]. And the length of polymethine chains is the key parameter in tuning the excitation/emission profile of the dye. For instance, the three methine protons named as Cy3 emits visible light, the five methine protons named as Cy5 emits near-red light, and the seven ones named as Cy7 emits in far-red region [15]. Most of these cyanine dyes performed narrow absorption bands, high extinction coefficients (greater than 200 000 $M^{-1}$ $cm^{-1}$), and weak fluorescers with quantum yields (1–18%). The in-depth understanding reveals that the extension of the polymethine bridge presents red shifts emission and, however, lower quantum yields [16–18]. While, the addition of cyclohexenyl moiety at the center of the polymethine bridge or complexation with proteins is benefit for enhancement of quantum yield and stability of these dyes attributing to "rigidizing" the backbone [18, 19]. Otherwise, the same purpose can either be reached by substituting various electrons donating at the N position on the 3H-indolenine rings enhancing photostability, while withdrawing groups in the same position usually decreased stability [20]. For instance, Peng et al. synthesized new heptamethine cyanine dyes by modifying the central chlorocyclohexenyl group resulting in larger blueshifts (>140 nm), lower molar extinction but higher fluorescence quantum yields [21]. On the contrary, Lee et al. react the same group to dyes by adding a carboxyl containing moiety to the central cyclohexenyl group. Although the newly developed dyes greatly improved the water solubility and offering a bioconjugation site, the absorption and emission of the dyes are redshifted with lower quantum yields [6].

### 2.2.2 Aggregates of Cyanine Dyes

For organic materials, construction of supramolecular J-aggregates is aroused significant interest since the germination of supramolecular chemistry in the 1930s [14]. In the dye chemistry, this discovery represents one of the most significant milestones in the history. The terms of Scheibe aggregates or Jelley aggregates (J-aggregates) are named according to their inventors. The definitions of the above term are bathochromically shifted (shifted to a longer wavelength) to the narrow absorption band of dye aggregates respecting to monomer absorption band and the nearly resonant fluorescence (very small Stokes shift) with narrow band [22–25]. On the contrary, the characters of hypsochromic aggregates (H-aggregates) are hypsochromically shifted (shifted to a shorter wavelength) to the absorption band respecting to monomer band and mostly in the case of low or no fluorescence [23]. Over 70 year ago, the J-aggregates inventors (Scheibe et al. [26] and Jelley [27]) independently observed the J-aggregation behavior of pseudo iso-cyanine chloride (1,1'-diethyl-2,2'-cyanine chloride or PIC chloride, as shown in Fig. 2.1) in aqueous solutions. They found that the absorption maximum was redshifted to $\tilde{v} = 17500$ cm$^{-1}$ ($\lambda_{max} = 571$ nm) upon increasing the concentration above $10^3$ M in water than other solvents, such as ethanol. With higher concentration ($10^2$ M), the absorption band became intense and sharp and hence large deviations from Lambert–Beer law (Fig. 2.1) [28]. In addition, the sodium chloride added into water resulted in similar phenomenon to increase the concentration of this dye. The characteristic features of the new absorption band are that the sharpness and high absorption coefficient. Meantime, the dyes exhibit strong fluorescence with small Stokes shift. Time moved to 1937, Scheibe et al. had already correctly interpreted the aggregation behavior of this dye, which attributed to desolvation upon supramolecular polymerization. The absorption spectrum was changed due to the adjacent molecules according to the "vicinity effect" [26]. In contrast, Jelley observed that this dyes showed a rapid precipitation from ethanol solution upon addition of nonpolar solvents or sodium chloride aqueous solution. The high fluorescence was lost as the dye passes into crystalline state [27]. Both two inventors found that the spectral change was able to reverse upon heating and cooling, which indicated the aggregation of the dye molecules.

**Fig. 2.1** Absorption spectra of PIC chloride aggregates in water (—) and its monomers in ethanol (- - - -), as well as the structure of PIC chloride (1)

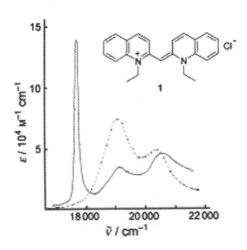

Although the J-aggregation of cyanine dyes is particular interest, the aggregation process from individual molecules to colloidal particles is poorly controlled. Herz presented J-aggregation of imidacarbocyanine molecules as an example to investigate the equilibria between the monomeric and aggregated forms of dyes according to the analysis of absorption spectra at different concentrations (Fig. 2.2) [23]. Based on the law of mass action, the equilibrium between monomers and aggregates can be written as follows [23]:

$$n \, c_M \rightleftharpoons c_J, \tag{1}$$

$$K = \frac{C_J}{C_M^n}$$

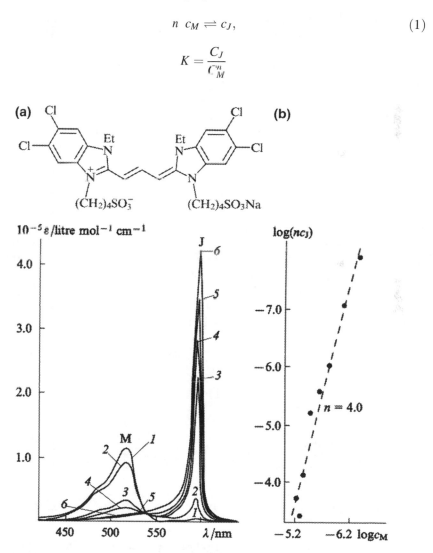

**Fig. 2.2** Dependence of absorption of benzimidacarbocyanine at 25 °C on its concentration in a $10^{-3}$ M NaOH aqueous solution. The concentration of the solution/mol $l^{-1}$: (1) $0.5 \times 10^{-6}$, (2) $1.0 \times 10^{-6}$, (3) $5.0 \times 10^{-6}$, (4) $1.0 \times 10^{-5}$, (5) $1.0 \times 10^{-4}$, (6) $4.0 \times 10^{-4}$; $C_M$ and $C_J$ are the concentrations of the monomeric and aggregated forms of the dye, respectively

$$\log(nc_J) = \log(nK) + \text{n} \log c_M, \tag{2}$$

where n is the number of monomers in J-aggregate, and $K$ is the equilibrium constant.

The relationship between $\log(nc_J)$ and logcM is plotted in linear with a slope of n. The change in the free energy of aggregation is determined from the temperature dependence of $K$.

The dyes of imidacarbocyanine in an aqueous electrolyte solution appeared n = 4 [23]. Herz hypothesized that the colloidal particles of dye molecules can be formed under experimental conditions. The n = 4 corresponded to the single spectroscopic unit rather than to the size of J-aggregates, because the same value of n was obtained by both J-aggregates of cationic naphthothiacarbocyanine [29] in aqueous solution and 1,10-diethyl-2,20-cyanine [30]. In some other studies, small values of n were determined from spectroscopic data [23]. It should be noted that irreversibility of the equilibrium in monomeric and J-aggregates undetermined the number of monomers in aggregates and the thermodynamic parameters of aggregation. The fact was that particles with different size gave similar absorption maxima. Thus, identification of small J-aggregated states or colloidal particles by spectroscopic methods is complicated. The reversibility of the equilibrium and colloidal stability in time in solution are the main criteria for determining the state of dyes [31]. Since then, the growing interest in J-aggregates of cyanine dye propelled the investigation of optical, photophysical, and structural properties of enormous an number of cyanine dyes [14]. The structure and morphology of J-aggregates of cyanine dyes were elucidated over the past few decades.

As two of the most widely known research amphiphilic cyanine dyes, C8O3 and its sulfo analogue C8S3 form, tubular aggregates are based on self-assembly. During the study of C8O3, von Berlepsch et al. gained the firsthand evidence which is the first time of tubular aggregates by cryo-TEM [32]. The structures resembling the rope were observed that are made up by bundles of a whole different number of

**(a)**

**(b)**

**(c)**

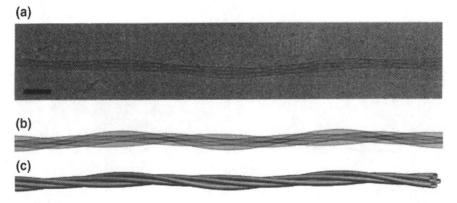

**Fig. 2.3** **a** Cryo-TEM image of a single quadruple helix. (Bar) 50 nm. **b** Simulated projection image of this helix. **c** Corresponding three-dimensional view

tubules. With the help of computer simulation, the structures are shown in Fig. 2.3. As shown in the picture, the hollow cylinders are packed on a triangular lattice, and at the same time, they are intertwined. The outer diameter of the tubules is 10 nm, and the length of the ropelike bundles is more than hundreds of micrometers. The thickness of the tubules' wall was about 4 nm, which led to the deduction that similar to a lipid bilayer, the wall is a bilayer structure of the amphiphilic dyes.

There are three forces to control the particular structure dispersion forces, hydrophobic forces, and electrostatic forces. The highly delocalized π-electron systems of the molecules will cause the dispersion forces. Hydrophobic forces will form if 1,1'-dialkyl substituents occur. The delocalized positive charge of the chromophore and ionized 3,3'-bis(ω-acidoalkyl) groups, and hydrogen bonds between the anionic acid groups will cause electrostatic forces. Up to now, it is not yet well understood which one is the most sensitive force among these three. These dye systems have an ability to form chiral J-aggregates from achiral molecules. And the symmetry which is between left- and right-handed species can be spontaneously broken.

Compared with those of C8O3, the tubules formed by self-assembly of the sulfobutyl-substituted dye C8S3 can be well separated, and only a few bundles of tubes are formed [33, 34]. The solvent composition and preparation conditions are the two factors to control the C8S3 tubes' diameter. In pure water, the diameter will be 16 nm and in solutions which contain more than 18 vol% methanol will be 13 nm [34]. No matter which cases, the thickness of the tubes' walls was observed to be on the order of 4 nm, which can be thought as an evidence that it is a bilayer wall, the same with C8O3. The double-walled tubular structure can be confirmed by a high-resolution transmission contrast image of C8S3 aggregates which is obtained by cryo-TEM. Fig. 2.4 shows a diagrammatic model of the double-walled tubular J-aggregates of C8S3. Such J-aggregates could be fixed on solid surfaces, which enabled the recording of high-resolution images of the topography and measurement of the exciton fluorescence of individual J-aggregates by polarization-resolved near-field scanning optical microscopy (NSOM) [14, 32, 33].

In this aspect, the first reported spectroelectrochemical studies of C8S3 cyanine dye double-walled tubular J-aggregates by Stevenson's groups fixed at an unfixed indium tin oxide (ITO) electrode surface. With the previous electrochemical studies of both solution-phase J-aggregates and cyanine dye monomers, the J-aggregates' permanent redox behavior has been found. The redox behavior is similar to the studies. The scan rate and pH both are the most important factors for the reversibility of J-aggregate oxidation. It suggests that both electrochemical and chemical steps are occurring in an overall electrochemical–chemical–electrochemical step (ECE) process mechanism (Fig. 2.5). Particularly, the original oxidation which both irreversible dimerization and dehydrogenation chemical steps tend to follow is the reason for the formation of a dehydrogenated dimer oxidation product. Both electrochemical analysis and in situ AFM measurements indicated that this dehydrogenated dimer product has its own unique redox activity. And that this product is most likely surface-confined has also been indicated. The studies about spectroelectrochemical show that the oxidation of J-aggregate is mostly

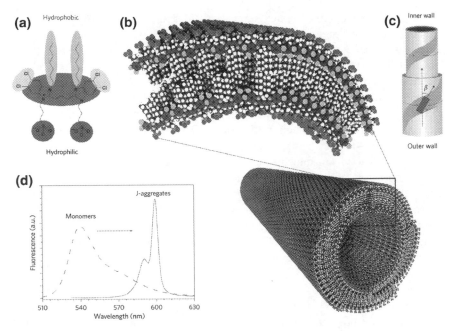

**Fig. 2.4** Cylindrical chiral double-walled nanotubular J-aggregates of an amphiphilic cyanine dye molecule. **a** 3,3'-bis(2-sulphopropyl)-5,5',6,6'- tetrachloro-1,1'-dioctylbenzimidacarbocyanine (C8S3) monomer. **b** Schematic of the self-assembled nanotube showing the double-walled structure with the alkyl chains at the interior of the bilayer. **c** Schematic showing the orientation, **b** of the transition dipole of the monomer relative to the long axis of the nanotube. The gray band demonstrates how the monomers wrap around the aggregate in both the inner and outer walls. **d** Fluorescence spectra of the monomer's solution and the aggregate's solution showing the narrowed and redshifted transitions typical for J-aggregates. The two main aggregate fluorescence peaks can be assigned to separate exciton transitions for the inner and outer walls

limited by the outer wall of the tubular structure, and compared to the outer wall, the inner wall has less oxidation than the outer wall. The excellent correlation between J-aggregate redox potentials and spectroelectrochemical data allows us to understand photo- and electron transfer processes in such cyanine dye J-aggregate systems [35].

The 2-(3-cyano-4,5,5-trimethyl-5H-furan-2-ylidene) malononitrile (TCF) was used as end groups for anionic cyanine which has high electronic delocalization, and N-ethyl-2-methyl-quinolinium was used as end groups for cationic cyanine by Jen's group. As an efficient electron acceptor for dipolar chromophores, the TCF end group, which has strong intramolecular charge transfer properties, was developed originally [36].

In a word, by designing the rational molecular, complementary salts of quinolinium- and TCF-cyanine have been used to adjust the molecular packing of heptamethine dyes in the solid state at the same time maintaining their highly polarizable and symmetric nature. We substantiated that famous J-aggregating

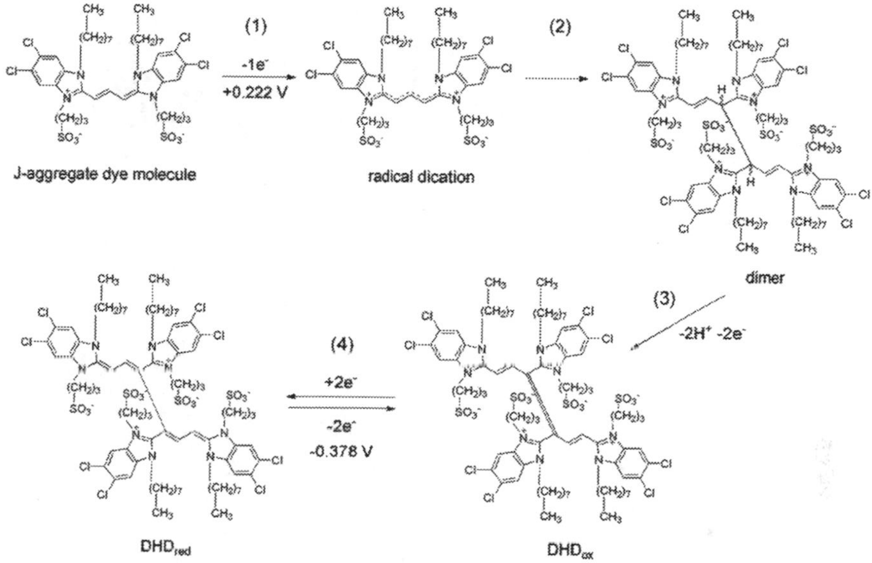

**Fig. 2.5** Depiction of the overall ECE mechanism for C8S3 J-aggregate oxidation and subsequent dehydrogenated dimer (DHD) formation

cationic cyanine can be used as structure-directing countercations for anionic cyanine. The most typical material is PIC. Our results show that when the phenyl substitution is at the center of conjugation backbone, it can suppress the potential H-type aggregation efficiently and strengthen the stability of J-type supramolecular aggregate structures. The first crystal structure of a complementary reported by us has a highly polarizable heptamethine backbone. It shows that both cationic cyanine and anionic cyanine in the complementary salt form a cooperative self-assembly. And it leads to a highly ordered J-aggregate-like molecular packing. This study designs a useful strategy to modify the aggregation and ion-pairing properties of highly polarizable cyanine with long conjugation length. Such rational design can give us the testification that the cyanine dyes have highly beneficial applications in biomedicine, optics, and other technologies [37].

The previous work developed a carrier system that consists of water-soluble neutral poly(phenylacetylene) and bore amphiphilic aza-18-crown-6 ether pendants (poly-1) to encapsulate the anchiral cyanine dye: 3,3-diethyloxadicarbocyanine iodide (O-5). In acidic water the dyes formed supramolecular J-aggregates inside the hydrophobic cavities of carriers. The chiral amino acids, such as L-or D-tryptophan (Trp), are capable of interacting with exterior crown ether pendants based on specific host–guest interactions. The host polymer predominantly formed one-hand helical formation. In the helical cavity of the polymer, the encapsulated cyanine aggregates generated optical activity, which was demonstrated by induced circular dichroism (ICD) in the achiral cyanine chromophore region. Meantime, the supramolecular chirality induced by the cyanine aggregated exhibited "memorized"

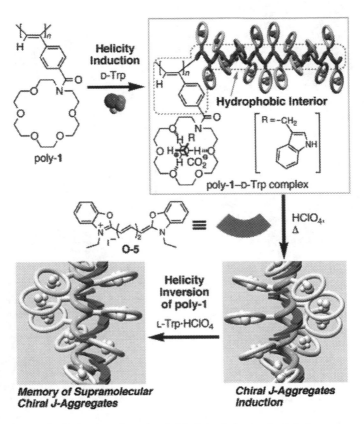

**Fig. 2.6** Schematic illustration of one-handed helicity induction in poly-1 upon complexation with D-Trp, subsequent supramolecular helical aggregate formation of achiral O-5 within the helical cavity of poly-1 in acidic water, and memory of the supramolecular chirality after helicity inversion of the poly-1 by excess L-Trp. The helix senses of poly-1 are assigned on the basis of the Cotton effect signs of the ICDs of analogous helical polyacetylenes and their helical senses determined by atomic force microscopy (AFM) observations

ability, which displayed opposite configuration after inversion of the templet polymer helicity (Fig. 2.6) [38].

The smallest possible cyanine dye aggregate is a dimer, which as a simple fact was significant in helping us determine how cyanine dye aggregates interacted with the DNA template. By using DNA templets with appropriate length and sequence, a helical aggregated cyanine dye dimer can be assembled literally one dimer at a time. This allows the photophysical and spectroscopic properties to be studied at each step during the process. Other dye aggregates are difficult to study because the number of dye molecules incorporated into the supramolecular structure is little and has no control. It presents a molecular model of three DiSC2(5) dimers that are assembled in the minor groove of DNA, which reflects the intimate relatives of the dye molecules and the DNA template. From this model, while the ethyl substituents project out of the groove, the exchange process that the sulfur atoms on the dye project into the minor groove occurs at the same time. This orientation causes that the curvature of the dye more closely matches the curvature of the groove [39]

## 2.3   Biomedical Applications of Cyanine Dyes

### 2.3.1   Biosensors for Bioactive Molecules

In the biomedical applications, the advantages in fluorescence and exciton transport properties of cyanine dye aggregates have been developed. For example, the cyanine J-aggregates can be efficiently quenched by minute amounts of analytes for biosensing applications [40]. Moreover, the pronounced color changes upon aggregation/disaggregation, or structural reorganization have been utilized in biosensor for recognition of a variety of biomacromolecules, including DNA, [40] polypeptides, and polysaccharides. All of these applications benefit from the enhanced solubility in aqueous solution of cationic cyanine dyes.

According to the literature, they devised a molecular-pairing technique, by which the amino acids are converted into isoindole derivatives (isoindole-amino acids) by reacting with orthophthalaldehyde (OPA) and alkyl thiols. Then, 2-Mercaptoethanesulfonic acid (MES) was added into the reaction system and offered an anionic charge which was tetherd to the isoindole unit.

Introducing aromatic and anionic groups into cyanine molecules was expected to enhance interactions with cationic molecules. By mixing the amino acids, OPA, and MES in water, the isoindole-amino acids were immediately obtained. The results were confirmed by fluorescence detection and new appeared absorption band at 333 nm. In the case of isoindole-Lys, both two amino groups are converted into isoindole units. The conclusion was confirmed by twofold increase of absorbance compared to that of the other amino acids. Employed cyanine dye as a functional molecular counterpart, the addition of cyanine dye to aqueous isoindole-amino acids led to an immediate color change from pink to reddish pink or orange. This phenomenon was depended on the chemical structure of amino acids (Fig. 2.7). However, the color change cannot be observed in the mixture of various amino acids and cyanine dye in the absence of OPA and MES. Thus, the reason for the color change was original from the interactions between isoindole-amino acids and dyes. The tendency to form H-aggregates of dyes is highly dependent on the chemical structure of the used amino acids. For further usage, this extended molecular-pairing approach enables to amplify and translate the molecular information of biomolecular components into spectroscopic and morphological information. The extended application of this approach will be a wide combination of biofunctional molecules. Besides aggregation property of cyanine dyes themselves or their derivatives

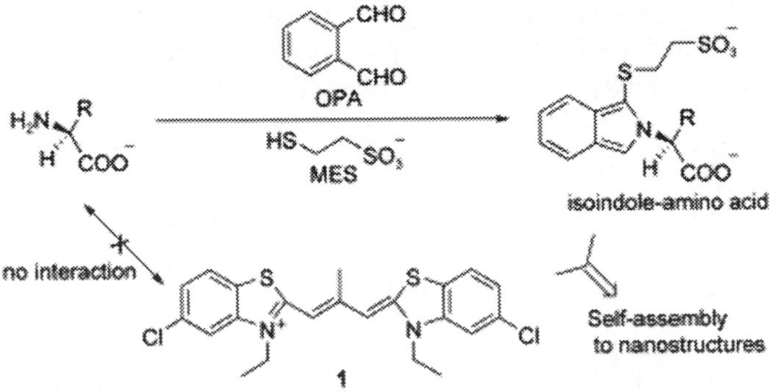

**Fig. 2.7** A schematic representation of the extended molecular pairing technique. Premodification of amino acids promotes the molecular association with the cationic cyanine dye, which develops into hierarchical nanostructures

provided characteristics, their co-assembly to small biomolecules and the assembly process may give more significant molecular information [41].

For instance, previous reports have provided J-aggregates from 3,30-disulfopropyl-5,50-dichlorothiacyanine (Tc) and 3,30-disulfobutyl-5,50-diphenyl-9-ethyloxacarbocyanine (Oc) in the aqueous solution containing NaCl, Mg(NO$_3$)$_2$, D/L-tartaric acids, asparagine, proline, DNA, and proteins (such as lysozyme, trypsin, RNase, and gelatin). In contrast to imidacarbocyanine and oxacarbocyanine, which are monomeric building blocks, thiacarbocyanine dyes exhibit specificity in their dimeric building blocks in J-aggregates. The J-aggregates formed in the presence of chiral additives are confirmed by sigmoidal kinetics with halftimes of 10–1000 s, resonance fluorescence, and large CD amplitudes being up to 2° for Tc. Generally, the induced CD signals of the J-aggregates of both dyes are bisignate and the sign corresponds to that of the additive. The chirality information transfers in the course of J-aggregation [42].

The supramolecular self-assembling cyanine is also used to investigate the DNA–drug interactions during high-throughput screening [43]. Supramolecular self-assembling cyanine and spermine binding to genomic DNA and the spermine competitively inhibited the self-assembling cyanine upon DNA scaffolds as reported signal by decreased fluorescence from the DNA–cyanine J-aggregate. The sequence of DNA exposure to cyanine or spermine was critical in determining the magnitude of inhibition. Moreover, methanol potentiated spermine inhibition by >tenfold. And they also demonstrated the reversibility of DNA–drug interactions by increasing concentration of cyanide that overcame spermine inhibition. Cyanine might be a safer alternative to the mutagenic ethidium bromide for investigating DNA–drug interactions.

Zhou's group [44] firstly demonstrated the supramolecular self-assembly of cyanines could be used to develop fluorescent enzymatic assays. They synthesized a

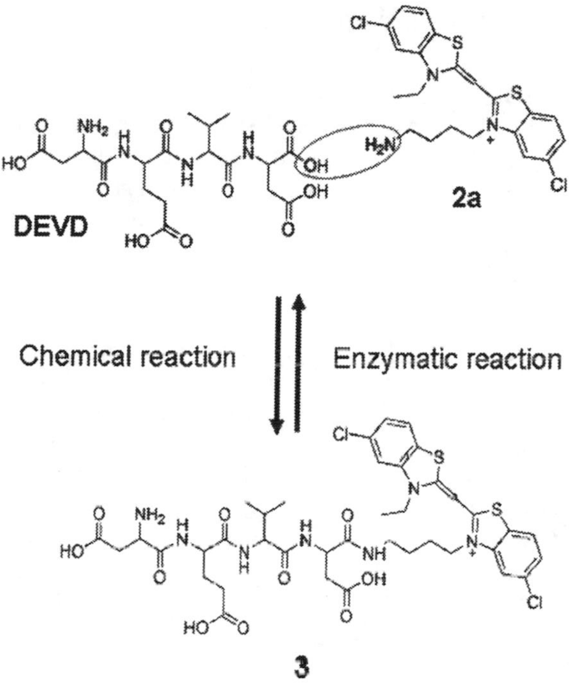

**Fig. 2.8** Schematics of DEVD cyanine 3 and hydrolysis by caspase-3 and caspase-7. This cartoon depicts the chemical coupling strategy of the deprotected DEVD tetrapeptide to aminocyanine 2a resulting in 3 (downward-pointing arrow). The enzymatic reaction (upward-pointing arrow) shows the hydrolysis of 3 through caspase-3 and caspase-7 proteolysis. The amino and carboxyl groups involved in the chemical and enzymatic reactions are circled. The schematic does not indicate any type of equilibrium existing between the chemical coupling reaction and the enzymatic reaction

covalent compound of the short-peptide Asp-Glu-Val-Asp (DEVD) and a cyanine (DEVD-cyanine).

The DEVD-cyanine due to its acknowledged sequence was specifically recognized and hydrolyzed by the caspase-3 and caspase-7 in 96- or 384-microwell plate reactions (Fig. 2.8). The catalytically released cyanine self-assembled on scaffolds of carboxymethylamylose (CMA) and carboxymethylcellulose (CMC) resulting in a *J*-aggregate displaying bright fluorescence at 470-nm emission wavelength. The fluorescence intensity increased with enzyme and substrate concentrations or reaction time and exhibited classical saturation profiles of a rectangular hyperbola. And the reaction kinetics was linear between 1 and 20 min and saturated at 60 min. The affinity constant (km) for DEVD-cyanine was ~ 23 μM, close to those of previously reported values for other DEVD substrates of caspase-3. So the method of cyanine dye supramolecular self-assembly can be used to detected protease activity.

## 2.3.2 Bioimaging and Diagnosis for Diseases

The modest photostability and ionic character of cyanine dye aggregates limit their scope for applications. Ongoing efforts in the research of cyanine dyes are focused on exploiting their value in biomedical applications [14]. Although amphiphilic cyanines possess many desirable properties, it has to be noted that these chromophores are without active biofunction. Recently, the challenge is developing the new cyanine dye aggregates with biofunction.

### 2.3.2.1 Cyanine Dye Nanovesicles for Tumor Bioimaging

Cyanine dye nanovesicles with defined structures and properties through precise self-assembly based on single building block would be a desirable contrast agent for bioimaging [45, 46] Meanwhile, it is still a challenge to obtain the nanovesicles derived from cyanine dye itself for bioimaging due to the certain water solubility (hydrophilicity). We have developed the $\pi$-conjugated bis(pyrene) building blocks, which prefer to self-aggregate in water due to large $\pi$-conjugation and strong hydrophobicity [47]. We introduced a bis(pyrene)-extended conjugated cyanine dye (BP-Cy) for constructing a well-organized nanovesicle with perfect chemical and photostability.

BP-Cy self-assembled into vesicular structure in DMSO/$H_2O$ mixed solution, which was imagined by transmission electron microscope (TEM) (Fig. 2.9). The thickness of the wall is $4.2 \pm 0.7$ nm, and the diameter of these nanovesicles is $46.6 \pm 6.0$ nm, which matches to the length of two molecules packed in a head-to-head mode [46]. Moreover, the BP-Cy nanovesicles were also observed by scanning electron microscope (SEM), which showed uniform nanoparticles in low magnification with a size of $59.9 \pm 9.8$ nm. The morphology of BP-Cy nanovesicles in situ was further observed by the atomic force microscope (AFM) instead of that in the dry state. The solution of BP-Cy nanovesicles was dropped on a mica substrate, and the flattened hollow nanovesicles with the diameter of $191.6 \pm 33.2$ and the height of $11.1 \pm 2.3$ were recorded (Fig. 2.9). The noteworthy shape changes on a hard substrate that further confirms the hollow structure of BP-Cy nanovesicles. To the best of our knowledge, the vesicular cyanine dye-based aggregates can only be observed in the presence of surfactant by template-induced structural transition effect [48]. We constructed cyanine dye-based vesicular nanostructures without any other inducing templates for the first time.

The NIR BP-Cy nanovesicles could be potential PA contrast agents in buffer solution, which showed high photostable. The almost stable UV–vis and PA signals of BP-Cy nanovesicles in water under continuous excitation at 790 nm for 180 min by UV2600 spectrometer and multispectral optoacoustic tomography (MOST) with 280 W/$cm^2$ for 60 min (Fig. 2.10), respectively confirmed the high stability. The self-assembled BP-Cy nanovesicles contribute to the high photostability due to the relative inert environment compared to small molecules in solution [49]. It is worth

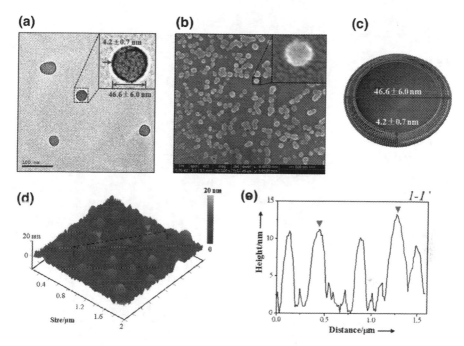

**Fig. 2.9** Morphologies of BP-Cy nanoaggregates. **a** TEM images of BP-Cy vesicles with negative staining. The expansions indicate the thickness of the bilayer membrane and the diameter of the vesicle in each case. **b** SEM images of BP-Cy nanoaggregates. **c** Schematic space-filling model for the vesicle. **d** AFM height images of BP-Cy nanoaggregates incubated on freshly cleaved mica surface at room temperature for 20 min. **e** Cross-sectional analysis along the dark line 1-1' in (**d**)

mentioning that the BP-Cy nanovesicles are much more stable than the commercial available ICG, which is regarded as the gold standard of PA contrast agent. Furthermore, photothermal experiment exhibited that the BP-Cy nanovesicles maintained 76.9% temperature increase of the third time compared to first time upon photoirradiation (Fig. 2.10). However, the parallel experimental value of ICG was 48.8%, indicating the much higher photostability of BP-Cy nanovesicles than ICG. On the basis of photostability results, the followed PA experiments were carried out on BP-Cy nanovesicles and ICG, as a reference. The PA spectrum of BP-Cy nanovesicles showed peak at 790 nm, which was identical to the UV–vis spectrum, indicating the solid correlation between absorption and PA intensity [50]. The photoacoustic intensity as a function of the molecular concentration of the ICG and BP-Cy nanovesicles exhibited a good linear relationship ($R^2$ = 0.98 (ICG) and 0.99 (BP-Cy nanovesicles, Fig. 2.10) [51]. The PA signals were ascribed to 16.2 times higher photothermal conversion efficiency. The BP-Cy nanovesicles are also very stable in various biological environments, e.g., PBS with different pH and cell culture media, which is more stable than the contrast agent ICG.

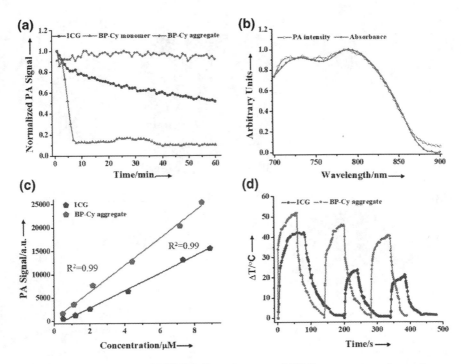

**Fig. 2.10 a** Photostability of ICG, BP-Cy monomer, and BP-Cy aggregate under continuous illumination by 790 nm. **b** The UV and PA spectra of BP-Cy aggregate in $H_2O$/DMSO (v/v = 99:1). **c** The PA signal produced by BP-Cy aggregate (red line) and ICG (blue line) was observed to be linearly dependent on the molecular concentration (R2 = 0.99). **d** The heating/cooling curves of BP-Cy and ICG solution for continuous 3 times. The energy input from lasers was 490 mW at 790 nm

With the high PA intensity and superb stability of BP-Cy nanovesicles, the BP-Cy nanovesicles were utilized as contrast agents for PA imaging to demonstrate the advantages. The long-term cell tracing was carried out with PA imaging by using human breast cancer MCF-7 cells [52, 53]. PA imaging of MCF-7 cells treated with BP-Cy nanovesicles and ICG, respectively, at different generations was recorded by the MOST. The results are that PA intensities of MCF-7 cells are 9.6 times higher than those of ICG-treated group at the first generation. Moreover, the PA intensity decreased to $\sim 3\%$ after continuous culture till the 6 generation and the parallel experimental till 3 generation treated with ICG. The results clearly indicated the noble cell tracing ability of BP-Cy nanovesicles over ICG dyes. The cell imaging based on BP-Cy nanovesicles with high stability for long time was carried out in agar-based phantom by using MCF-7 cells. The $T_{p1/2}$, which was defined as the time interval for PA signals decreased to half of the original ones, of BP-Cy nanovesicles incubated cells was longer than 5 d. As a reference, the $T_{p1/2}$ of ICG-treated cells was 0.5 d.

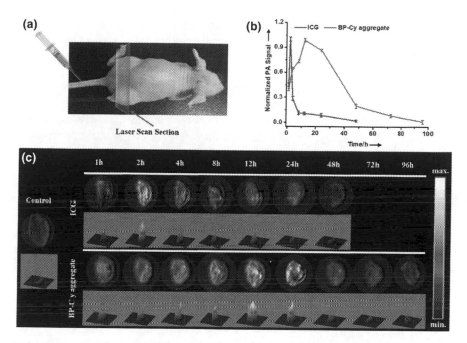

**Fig. 2.11** In vivo PA imaging of mice. **a** In vivo model mice for PA imaging injected with ICG and BP-Cy aggregate (20 μM, 200 μL) via tail vein and detected under laser irradiation (700–850 nm). **b** The quantification of normalized PA signals from the tumor injected with ICG (blue line) and BP-Cy aggregate (red line) for different time intervals and corresponding PA images of tumors (**c**)

To demonstrate the advantages of the BP-Cy nanovesicles as a high-performance PA contrast agent, we carried out the PA imaging of tumor-xenografted mice based on the BP-Cy nanovesicles (Fig. 2.11). For this, BP-Cy nanovesicles were administered systemically into MCF-7 tumor-bearing mice via tail vein injection. To our interest, the PA signals at tumor site were monitored and collected upon 700–850 nm excitations (Figs. 2.11) [54]. After 2-h administration, the PA signals were clearly observed in the tumor region, which was further increased from 2 to 12 h and reached a maximum at 12-h post-injection. After that, the PA signals decreased to a detectable level until 72 h. As a control group, tumor-xenografted mice were treated with ICG in the same condition, and the PA signals only retained for 12 h with the maxima at 2 h. The continuous PA intensity in vivo could be attributed to the high accumulation and long retention of BP-Cy nanovesicles in tumor compared with ICG molecules (Fig. 2.11) [55]. Importantly, high photo- and chemical stability in tumor site is necessary for long time in vivo imaging.

Owing to their intrinsic structural characteristic, the cyanine dyes preferentially self-assembled into lamellar or tubular superstructures. We tailored the p-scaffolds of cyanine dyes and for the first time reported their vesicular self-assembly without templates, to the best of our knowledge. Interestingly, compared with the monomer and analogue ICG molecules, the stable nanovesicles show significantly improved PA signal intensity and half-life in vitro and in vivo. We have constructed a supramolecular NIR-absorbing cyanine dye-based aggregate as a long-term PA imaging contrast agent. Our results demonstrated that supramolecular strategy can effectively improve the photophysical properties of building blocks, which are highly demanded for the bottom-up fabrication of functional materials and devices. We believe that this hollow nanostructure can be potentially employed as a platform for delivery of various bioactive molecules for extensive biomedical applications.

### 2.3.2.2  Cyanine Dye Nanoemitters for Cancer Diagnosis

Supramolecular assemblies have been considered as a powerful approach to fabricate artificial architectures for biomedical applications, such as bioimaging [56, 57], drug delivery [58], and tissue engineering [59–62]. However, for application in biological system, the interaction with the complex physiological environment always led such transformation uncontrollable. Thus, the ingenious design of the building blocks and the triggered factor for self-assembled nanomaterial fabrication are still now a challenging. We developed a new bioorthogonal reaction as an activator for converting the cyanine dye nanoemitters from "inert" to "active" states with binary fluorescence developed for tumor diagnostics in vivo. The new molecule included covalently two emissive cyanine (Cy) and bis(pyrene) (BP) motifs, which self-assembled into nanoemitters in "inert" states with quenched fluorescence due to aggregation-caused quenching (ACQ) [63, 64].

When meeting GSH, the cyanine (Cy)-based dye was selectively separated. The GSH was spontaneously reacted with Cy moiety through bioorthogonal reaction, and the rest of the moiety composting with hydrophobic BP consequently grow to larger aggregates accompanying dual-color fluorescence signals turned-on (Fig. 2.12). Especially, the enhancement of dual-color fluorescence exhibited 30-fold for GSH-labeled Cy at NIR region ($\lambda_{em}$ = 820 nm) and 36-fold for BP aggregates at 520 nm. This highly sensitive bioimaging system is successfully applied in vivo imaging of GSH with dual-wavelength emission. As a result, the self-assembled nanoemitters 1 spontaneously exhibits binary emissions for high-performance tumor imaging in vivo. We believe that this bioorthogonally deciphered strategy opens a new avenue for designing variable smart biomaterials or devices in biomedical applications.

### 2.3.2.3  Cyanine Dye Nanopaticles for Bacterial Bioimaging

A central problem in imaging bacterial infections is to develop targeting strategies that can deliver large quantities of imaging probes to bacteria. This has been challenging because most present imaging probes target the bacterial cell wall and cannot access the bacterial intracellular volume [65–68]. In contrast, we developed a new cyanine dye nanoparticle through specifically structure-recognition

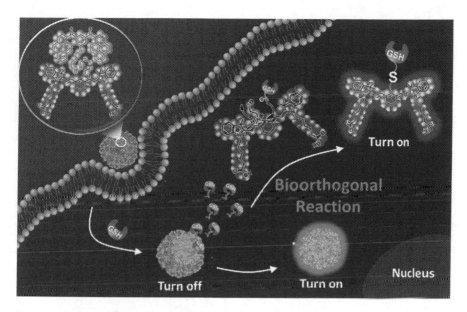

**Fig. 2.12** Schematic illustration of a new GSH-based bioorthogonal reaction that enables to decipher the nanometers of cyanine (Cy)-based dye (1) resulting in turn-on of dual-color fluorescence signals with a large shift excitation/emission profiles in living cells

peptidoglycan mechanism. This new nanoparticle is unprecedentedly sensitive and specific in bacterial infections imaging in vivo.

We introduced alkyl chain-modified N-alkyl side chains based on the heptamethine core for the construction of a well-organized NIR-absorbing nanoparticle 1. Molecule 1 shows a main absorption band at 700 nm and an additional less intense absorption band at 790 nm with the fluorescence dramatically quenched in $H_2O$. This result suggested molecule 1 changed to aggregated state in $H_2O$. We systematically studied the optical property of molecule 1 from monomeric state to aggregate state in DMSO gradually changed to $H_2O$. The absorption band becomes broader with a shoulder peak at a longer wavelength around 850 nm. Meanwhile, the fluorescence of molecule 1 was dramatically quenched upon addition of $H_2O$. The molecule 1 (10 mM) aggregates self-assembled into micelle structures in the DMSO/$H_2O$ mixed solution (v:v = 1:99), which was visualized using a transmission electron microscope (TEM). The results indicated that the nanoparticle 1 showed particulate structures with a size (49.1 ± 4.0 nm), which well corresponded to the results obtained by dynamic light scattering (DLS) measurement (59.8 ± 11.6 nm).

Peptidoglycan is a polymer consisting of sugars and amino acids that forms a mesh-like layer outside the plasma membrane of most bacteria [69, 70]. The peptide chain can be cross-linked to the peptide chain of another strand forming the 3D mesh-like layer [69]. The nanoparticle 1 can specifically recognize the peptidoglycan, then it disassembles into molecule 1 which blocks in hydrophobic cave of

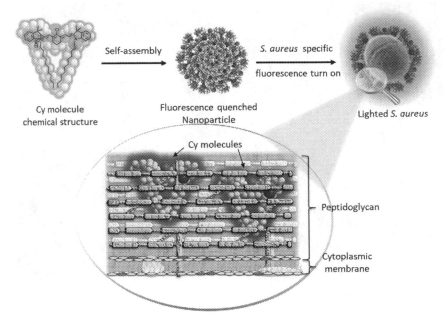

**Fig. 2.13** Schematic illustration of the mechanism that the cyanine dye nanoparticle specifically targets and detects infections of Gram-positive bacteria

peptidoglycan, turning-on red fluorescence. The molecule 1 instills durability 3D mesh-like layer of peptidoglycan that can limit the intermolecular interactions by altering the steric and/or ion-pair properties of the molecule 1 (Fig. 2.13).

The peptidoglycan layer is substantially thicker in Gram-positive bacteria (20 to 80 nm) than in Gram-negative bacteria (7 to 8 nm). Peptidoglycan forms around 90% of the dry weight of Gram-positive bacteria but only 10% of Gram-negative strains. Thus, peptidoglycan is the primary determinant of the characterization of bacteria as Gram-positive [69, 70]. So the nanoparticle 1 can image Gram-positive bacterial infections in vivo with unprecedented sensitivity and specificity.

The in vivo applicability of the nanoparticle 1 was subsequently tested in immunocompetent mice that had myositis in the right-hind limb induced by intramuscular injection of Gram-positive bacteria (*S. aureus* and *S. epidermidis*) and the control group of Gram-negative bacteria (*E. coli, S. marcescens* and *P. aeruginosa*). Two days after intramuscular inoculation with five bacteria respectively, mice were intravenously injected with the nanoparticle 1. After an additional 2 h, fluorescence imaging was performed with an IVIS Lumina II imaging system. The nanoparticle 1 yielded strong bacteria-specific signals in rats after intravenous administration. A strong NIR fluorescence (NIRF) signal, attributed to the nanoparticle 1, was observed for Gram-positive bacteria (*S. aureus* and

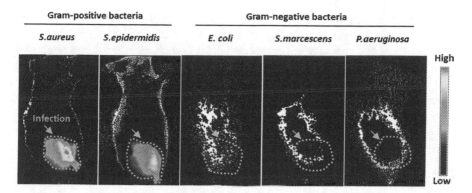

**Fig. 2.14** Fluorescence imaging of bacterial infection in vivo. Schematic illustration of imaging of infected sites in the leg muscles (injected with Gram-positive bacteria: *S. aureus* and *S. epidermidis,* and Gram-negative bacterial: *E. coli, S. marcescens* and *P. aeruginosa*)

*S. epidermidis*) (Fig. 2.14). In contrast, no NIR signal was observed for Gram-negative bacteria (*E. coli, S. marcescens* and *P. aeruginosa*) (Fig. 2.14). The heart, spleen, normal muscle, and bowel showed no fluorescence signal above background levels. The nanoparticle 1 was able to discriminate Gram-positive bacteria infection from Gram-negative bacteria-induced sterile inflammation in vivo.

Here, guided by recent advances in clinically relevant optical imaging technologies, we explore NIR cyanine dye nanoparticle to specifically target and detect infections caused by Gram-positive bacteria. We show that NIR optical imaging with a Gram-positive bacteria-specific NIR cyanine dye nanoparticle can detect both invasive infections real time in vivo. We conclude that NIR cyanine dye nanoparticle has a high potential for enhanced noninvasive diagnosis of infections with Gram-positive bacteria and is a promising candidate for early-phase clinical trials.

## 2.4 Summary and Outlook

From a supramolecular perspective, it is remarkable that the exact aggregate structure of the first *J*-aggregate, that is, PIC aggregate, is still a matter of debate 75 years after its discovery and will remain a topic of future research. For biomolecular science, water solubility is of relevance, and thus, it is not surprising that cationic cyanine dye aggregates were the dyes of choice for the development of novel fluorescent and colorimetric sensing schemes, for example, by the connection of cyanine dyes to peptides or by DNA-templated J-aggregation. The strong absorption of light and efficient exciton transport to analyte traps provide these dye aggregates with an ultrahigh sensitivity. The challenges are developing new biofunctional self-assembled supramolecular systems. The use of self-assembled biomaterials in drug delivery and cancer therapy could also benefit from improved nanostructural precision. In addition, the native matrix provides spatiotemporal control of bioactivity, chemical composition, and mechanical properties as tissues develop or cells mature. It is postulated that the controlled growth or templated assembly of such carriers should lead to more consistent performance in vivo. One vision for the future of supramolecular biomaterials would seek to more closely mimic both the structural and functional aspects of native extracellular matrix.

# References

1. Stephens DJ, Allan VJ (2003) Light microscopy techniques for live cell imaging. Science 300 (5616):82–86
2. Medintz IL, Uyeda HT, Goldman ER, Mattoussi H (2005) Quantum dot bioconjugates for imaging, labelling and sensing. Nat Mater 4(6):435–446
3. Lilley DMJ, Wilson TJ (2000) Fluorescence resonance energy transfer as a structural tool for nucleic acids. Curr Opin Chem Biol 4(5):507–517
4. Royer CA (2006) Probing protein folding and conformational transitions with fluorescence. Chem Rev 106(5):1769–1784
5. Stennett EMS, Ciuba MA, Levitus M (2014) Photophysical processes in single molecule organic fluorescent probes. Chem Soc Rev 43:1057–1075
6. Pansare VJ, Hejazi S, William JF, Robert KP (2012) Review of long-wavelength optical and NIR imaging materials: contrast agents, fluorophores, and multifunctional nano carriers. Chem Mater 24:812–827
7. Ghoroghchian PP, Frail PR, Susumu K, Park T-H, Wu SP, Uyeda HT, Hammer DA, Therien MJ (2005) Broad spectral domain fluorescence wavelength modulation of visible and near-infrared emissive polymersomes. J Am Chem Soc 127(44):15388–15390
8. Luo S, Zhang E, Su Y, Cheng T, Shi C (2011) A review of NIR dyes in cancer targeting and imaging. Biomaterials 32:7127–7138
9. Ogawa M, Kosaka N, Choyke PL, Kobayashi H (2009) In vivo molecular imaging of cancer with a quenching near infrared fluorescent probe using conjugates of monoclonal antibodies and indocyanine green. Cancer Res 69(4):1268–1272
10. Guo Z, Park S, Yoon J, Shin I (2014) Recent progress in the development of near-infrared fluorescent probes for bioimaging applications. Chem Soc Rev 43(1):16–29
11. Williams CG (2013) XXVI—researches on chinoline and its homologues. Trans Royal Soc Edinburgh 21(3):377–401
12. König W (1922) Über die Konstitution der Pinacyanole, ein Beitrag zur Chemie der Chinocyanine. Berichte der deutschen chemischen Gesellschaft (A and B Series) 55(9):3293–3313
13. König W (1926) Über den Begriff der Polymethinfarbstoffe und eine davon ableitbare allgemeine Farbstoff-Formel als Grundlage einer neuen Systematik der Farbenchemie. J für Praktische Chemie 112(1):1–36
14. Würthner F, Kaiser TE, Saha-Möller CR (2011) J-aggregates: from serendipitous discovery to supra-molecular engineering of functional dye materials. Angew Chem Int Edit 50:3376–3410
15. McRae EG, Kasha M (1958) Enhancement of phosphorescence ability upon aggregation of dye molecules. J Chem Phys 28(4):721–722
16. Jones RM, Bergstedt TS, Buscher CT, McBranch D, Whitten D (2001) Superquenching and its applications in j-aggregated cyanine polymers. Langmuir 17(9):2568–2571

17. Jones RM, Lu L, Helgeson R, Bergstedt TS, McBranch DW, Whitten DG (2001) Building highly sensitive dye assemblies for biosensing from molecular building blocks. Proc Natl Acad Sci 98(26):14769–14772

18. Baldwin RL (1996) How Hofmeister ion interactions affect protein stability. Biophys J 71 (4):2056–2063

19. Stoll RS, Severin N, Rabe JP, Hecht S (2006) Synthesis of a novel chiral squaraine dye and its unique aggregation behavior in solution and in self-assembled monolayers. Adv Mater 18 (10):1271–1275

20. Alex S, Basheer MC, Arun KT, Ramaiah D, Das S (2007) Aggregation properties of heavy atom substituted squaraine dyes: evidence for the formation of j-type dimer aggregates in aprotic solvents. J Chem Phys A 111(17):3226–3230

21. Wojtyk J, McKerrow A, Kazmaier P, Buncel E (1999) Quantitative investigations of the aggregation behaviour of hydrophobic anilino squaraine dyes through UV/vis spectroscopy and dynamic light scattering. Can J Chem 77(5–6):903–912

22. Egorov VV (2002) Nature of the optical transition in polymethine dyes and J-aggregates. J Chem Phys 116(7):3090–3103

23. Herz AH (1977) Aggregation of sensitizing dyes in solution and their adsorption onto silver halides. Adv Colloid Interface Sci 8(4):237–298

24. Fidder H, Knoester J, Wiersma DA (1991) Optical properties of disordered molecular aggregates: a numerical study. J Chem Phys 95(11):7880–7890

25. Rousseau E, Koetse MM, Van der Auweraer M, De Schryver FC (2002) Comparison between J-aggregates in a self-assembled multilayer and polymer-bound J-aggregates in solution: a steady-state and time-resolved spectroscopic study. Photochem Photobiol Sci 1(6):395–406

26. Scheibe G, Kandler L, Ecker H (1937) Polymerisation und polymere Adsorption als Ursache neuartiger Absorptionsbanden von organischen Farbstoffen. Naturwissenschaften 25 (5).75–75

27. Jelley EE (1937) Molecular, nematic and crystal states of I, I'-diethyl-ψ-cyanine chloride. Nature 139:631–632

28. Kopainsky B, Hallermeier JK, Kaiser W (1981) The first step of aggregation of pic: the dimerization. Chem Phys Lett 83(3):498–502

29. Ballard RE, Gardner BJ (1971) The J-band of 1,1[prime or minute]-diethyl-9-methyl-4,5;4 [prime or minute],5[prime or minute]-dibenzthiacarbocyanine chloride. J Chem Soc B Phys Org (0):736–738

30. Michelbacher E (1969) Abklingzeitmessungen an Pseudoisocyanindiäthylchlorid mit einem Phasenfluorometer mit 200 MHz-Lichtmodulation. Zeitschrift für Naturforschung A, vol 24

31. Shapiro BI (2006) Molecular assemblies of polymethine dyes. Russ Chem Rev 75:433–456

32. Hv Berlepsch, Bottcher C, Ouart A, Burger C, Dahne S, Kirstein S (2000) Supramolecular structures of j-aggregates of Carbocyanine dyes in solution. J Phys Chem B 104:5255–5262

33. Eisele DM, Knoester J, Kirstein S, Rabe JP, Vanden Bout DA (2009) Uniform exciton fluorescence from individual molecular nanotubes immobilized on solid substrates. Nat Nanotechnol 4(10):658–663

34. Didraga C, Pugžlys A, Hania PR, von Berlepsch H, Duppen K, Knoester J (2004) Structure, spectroscopy, and microscopic model of tubular carbocyanine dye aggregates. J Chem Phys B 108(39):14976–14985

35. Lyon JL, DrM Eisele, Kirstein S, Rabe JP, Bout DAV, Stevenson KJ (2008) Spectroelectrochemical investigation of double-walled tubular j-aggregates of amphiphilic cyanine dyes. J Phys Chem C 112:1260–1268

36. Liu S, Haller MA, Ma H, Dalton LR, Jang SH, Jen AKY (2003) Focused microwave-assisted synthesis of 2,5-dihydrofuran derivatives as electron acceptors for highly efficient nonlinear optical chromophores. Adv Mater 15(7–8):603–607

37. Za Li, Mukhopadhyay S, Jang S-H, Brédas J-L, Jen AK-Y (2015) Supramolecular assembly of complementary Cyanine salt j-aggregates. J Am Chem Soc 137:11920–11923

38. Miyagawa T, Yamamoto M, Muraki R, Onouchi H, Yashima E (2007) Supramolecular helical assembly of an Achiral Cyanine dye in an induced helical amphiphilic Poly(phenylacetylene) interior in water. J Am Chem Soc 129:3676–3682
39. Hannah KC, Armitage BA (2004) DNA-templated assembly of helical Cyanine dye aggregates: a supramolecular chain polymerization. Acc Chem Res 37:845–853
40. Whitten DG, Achyuthan KE, Lopez GP, Kim O-K (2006) Cooperative self-assembly of cyanines on carboxymethylamylose and other anionic scaffolds as tools for fluorescence-based biochemical sensing. Pure Appl Chem 78:2313–2323
41. Shiraki T, M-a Morikawa, Kimizuka N (2008) Amplification of molecular information through self-assembly: nanofibers formed from amino acids and cyanine dyes by extended molecular pairing. Angew Chem-Int Edit 120:112–114
42. Slavnova TD, Görner H, Chibisov AK (2011) Cyanine-based j-aggregates as a chirality-sensing supramolecular system. J Phys Chem B 115:3379–3384
43. Achyuthan KE, Whitten DG, Branch DW (2010) Supramolecular self-assembling Cyanine as an alternative to ethidium bromide displacement in DNA–drug model interactions during high throughput screening. Anal Sci 26:55–61
44. Zhou Z, Tang Y, Whitten DG, Achyuthan KE (2009) New high-throughput screening protease assay based upon supramolecular self-assembly. ACS Appl Mater Interfaces 1 (1):162–170
45. Huynh E, Lovell JF, Helfield BL, Jeon M, Kim C, Goertz DE, Wilson BC, Zheng G (2012) Porphyrin shell microbubbles with intrinsic ultrasound and photoacoustic properties. J Am Chem Soc 134(40):16464–16467
46. Hong G, Zou Y, Antaris AL, Diao S, Wu D, Cheng K, Zhang X, Chen C, Liu B, He Y, Wu JZ, Yuan J, Zhang B, Tao Z, Fukunaga C, Dai H (2014) Ultrafast fluorescence imaging in vivo with conjugated polymer fluorophores in the second near-infrared window. Nat Commun 5(4206):1–8
47. Wang L, Li W, Lu J, Zhao Y-X, Fan G, Zhang J-P, Wang H (2013) Supramolecular nano-aggregates based on Bis(Pyrene) derivatives for lysosome-targeted cell imaging. J Phys Chem C 117(50):26811–26820
48. Barni E, Savarino P, Pelizzetti E, Rothenberger G (1981) Synthesis, surface activity and micelle formation of novel Cyanine dyes. Helv Chim Acta 64(6):1943–1948
49. Guo Z, Park S, Yoon J, Shin I (2014) Recent progress in the development of near-infrared fluorescent probes for bioimaging applications. Chem Soc Rev 43(1):16–29
50. Krumholz A, Shcherbakova DM, Xia J, Wang LV, Verkhusha VV (2014) Multicontrast photoacoustic in vivo imaging using near-infrared fluorescent proteins. Sci Rep 4(3939):1–7
51. Wang K-R, An H-W, Rong R-X, Cao Z-R, Li X-L (2014) Synthesis of biocompatible Glycodendrimer based on fluorescent perylene bisimides and its bioimaging. Macromol Rapid Commun 35(7):727–734
52. Li K, Qin W, Ding D, Tomczak N, Geng J, Liu R, Liu J, Zhang X, Liu H, Liu B, Tang BZ (2013) Photostable fluorescent organic dots with aggregation-induced emission (AIE dots) for noninvasive long-term cell tracing. Sci Rep 3(1150):1–10
53. Wang Z, Chen S, Lam JWY, Qin W, Kwok RTK, Xie N, Hu Q, Tang BZ (2013) Long-term fluorescent cellular tracing by the aggregates of aie bioconjugates. J Am Chem Soc 135 (22):8238–8245
54. Zhang D, Qiao Z-Y, Mayerhoeffer U, Spenst P, Li X-J, Würthner F, Wang H (2014) Nano-confined Squaraine dye assemblies: new photoacoustic and near-infrared fluorescence dual-modular imaging probes in vivo. Bioconjug Chem 25(11):2021–2029
55. de la Zerda A, Liu Z, Bodapati S, Teed R, Vaithilingam S, Khuri-Yakub BT, Chen X, Dai H, Gambhir SS (2010) Ultrahigh sensitivity carbon nanotube agents for photoacoustic molecular imaging in living mice. Nano Lett 10(6):2168–2172
56. Ye D, Liang G, Ma ML, Rao J (2011) Controlling intracellular macrocyclization for the imaging of protease activity. Angew Chem Int Edit 50(10):2275–2279

57. Wang H, Wang S, Su H, Chen KJ, Armijo AL, Lin WY, Wang Y, Sun J, Kamei K, Czernin J, Radu CG, Tseng HR (2009) A supramolecular approach for preparation of size-controlled nanoparticles. Angew Chem Int Edit 48(24):4344–4348

58. Yao Y, Xue M, Chen J, Zhang M, Huang F (2012) An amphiphilic pillar [5] arene: synthesis, controllable self-assembly in water, and application in Calcein release and TNT adsorption. J Am Chem Soc 134(38):15712–15715

59. Williams RJ, Smith AM, Collins R, Hodson N, Das AK, Ulijn RV (2009) Enzyme-assisted self-assembly under thermodynamic control. Nat Nanotechnol 4(1):19–24

60. Li LL, Qi GB, Yu F, Liu SJ, Wang H (2015) An adaptive biointerface from self-assembled functional peptides for tissue engineering. Adv Mater 27(20):3181–3188

61. Zhang S (2003) Fabrication of novel biomaterials through molecular self-assembly. Nat Biotechnol 21(10):1171–1178

62. Zhang S (2003) Fabrication of novel biomaterials through molecular self-assembly. Nat Biotechnol 21(10):1171–1178

63. An HW, Qiao SL, Hou CY, Lin YX, Li LL, Xie HY, Wang Y, Wang L, Wang H (2015) Self-assembled NIR nanovesicles for long-term photoacoustic imaging in vivo. Chem Commun 51(70):13488–13491

64. van Bünau G, Birks JB (1970) Photophysics of aromatic molecules. Wiley, London 1970. 704 Seiten. Preis: 210s. Berichte der Bunsengesellschaft für physikalische Chemie 74 (12):1294–1295

65. Leevy WM, Gammon ST, Jiang H, Johnson JR, Maxwe DJ, Jackson EN, Marquez M, Piwnica Worms D, Smith BD (2006) Optical imaging of bacterial infection in living mice using a fluorescent near-infrared molecular probe. J Am Chem Soc 128:16476 16477

66. Smith BA, Akers WJ, Leevy WM, Lampkins AJ, Xiao S, Wolter W, Suckow MA, Achilefu S, Smith BD (2010) Optical imaging of mammary and prostate tumors in living animals using a synthetic near infrared Zinc(II)-Dipicolylamine probe for anionic cell surfaces. J Am Chem Soc 132(1):67–69

67. Welling MM, Paulusma-Annema A, Balter HS, Pauwels EKJ, Nibbering PH (2000) Technetium-99 m labelled antimicrobial peptides discriminate between bacterial infections and sterile inflammations. Eur J Nucl Med 27(3):292–301

68. Ning X, Lee S, Wang Z, Kim D, Stubblefield B, Gilbert E, Murthy N (2011) Maltodextrin-based imaging probes detect bacteria in vivo with high sensitivity and specificit. Nat Mater 10:602–607

69. Schleifer KH, Kandler O (1972) Peptidoglycan types of bacterial cell walls and their taxonomic implications. Bacteriol Rev, 407–477

70. Weidenmaier C, Peschel A (2008) Teichoic acids and related cell-wall glycopolymers in gram-positive physiology and host interactions. Nat Rev Microbiol 6(4):276 287

# Chapter 3
# Self-assembled Nanomaterials for Bacterial Infection Diagnosis and Therapy

Li-Li Li

**Abstract** Self-assembled nanomaterials are composed of building blocks through non-covalent interaction and spontaneously arranged into well-ordered nanostructures with defined functions. The well-organized arrangement and the two-/three-dimensional nanostructure of the architectures endow the nanomaterials abundant excellent biofunctions for bacterial infection detection and therapy applications. Beyond nature-inspired sources, the hybrid artificial nanomaterials including inorganic nanoparticles, nanosized small synthetic molecules assemblies, self-assembled multilayer polymers are reviewed in this chapter. In addition, the design concept, assembled driving forces, nanostructural effect, antimicrobial mechanism, detection methods are also discussed and summarized. As the promising field, in vivo self-assembled nanomaterials with specific stimuli-responsiveness and surprised biofunctions are also included in this chapter to explore and fabricate fascinated self-assembled nanomaterials.

## 3.1 Introductions

Self-assembled nanomaterials represent a classic of non-covalent interactions induced self-assemblies with well organized nanostructures. The self-assembled process can be divided into kinetic- and thermodynamic-controlled assembly in a spatial and temporal manner. Inspired by nature, the self-assembly in organism, which ubiquitously built biofunctional networks, plays a significant role during the bioregulation process. It can be said that bio-system is constructed and evolved by self-assembly-induced biofunctional innovation. Hence, people are trying to reveal and imitate natural self-assemblies, spontaneously devoting themselves to understand the mechanisms. Beyond nature heuristic, artificially designed self-assembled

L.-L. Li (✉)
CAS Center for Excellence in Nanoscience, CAS Key Laboratory for Biomedical Effects
of Nanomaterials and Nanosafety, National Center for Nanoscience and Technology
(NCNST), No. 11, Beiyitiao, Zhongguancun, Beijing 100190, China
e-mail: lill@nanoctr.cn

© Springer Nature Singapore Pte Ltd. 2018
H. Wang and L.-L. Li (eds.), *In Vivo Self-Assembly Nanotechnology for Biomedical
Applications*, Nanomedicine and Nanotoxicology,
https://doi.org/10.1007/978-981-10-6913-0_3

nanomaterials also obtained the considerable development during the past decades. All the efforts contributed a lot in bacterial infection diagnosis and therapy in the purpose of early detection and less drug resistance [1].

With the development of the synthetic chemistry and material science, the molecular building blocks, which can be considered for antimicrobial or detection agents, are promising with peptides, nucleotides, biopolymers, proteins, photothermal/photodynamic agent, or even inorganic nanomaterials. The endless imagination drives the development of the design of the novel self-assembled nanomaterials with well-controlled nanostructures and fantastic biofunctions. Surprisingly from natural antimicrobial sources, such as antimicrobial peptides (AMPs) [2–4], cationic natural polymers [5, 6], natural photothermal/photodynamic molecules [7–9] , sophisticated molecular structures are analyzed and modified as building blocks for self-assembly with enhanced antimicrobial efficiency. Besides direct bacteria killing, self-assembled materials with well-ordered assembly surface multilayers nanostructures exhibited an excellent antifouling and anti-biofilm property, which indirectly contribute to bacterial inhibition as a surface defense [10, 11]. Of course, other than organic self-assembled nanomaterials, some hybrid inorganic nanomaterials present a strong growth tendency due to their unique antimicrobial mechanisms with fewer side effects through a self-assembled modification [12]. To defense human bacterial infection, antibiotic plays an irreplaceable role in history; however, the server antibiotic-resistance risk highly reduced the lifetime of newly developed antibiotic [13]. Here, some chemist gives us an enlightenment that self-assembly of such modified small antibiotic molecules might improve the antimicrobial efficiency and even change their antimicrobial mechanism [14].

Besides antimicrobial approach, the bacterial detection also contributed to solve the drug resistance risk by early, sensitive, and specific bacterial detection and identification leading to the significant guiding for rational use of drugs [15]. Therein, self-assembled nanomaterials endowed stimuli-responsiveness capability and showed wildly spread application in highly sensitive and specific bacterial detection [16, 17] or bacterial infection bioimaging [18, 19].

In this chapter, we will introduce various self-assembled nanomaterials investigated in the latest 5 years used for bacterial infection diagnosis and therapy, including nature sourced and artificially made building blocks. The design concept, assembled driving forces, nanostructural effect, antimicrobial mechanism, detection methods will be reviewed in the following part. Especially, we will outlook a recently improved promising field: in vivo self-assembled nanomaterials to better understand the development of self-assembly.

## 3.2  Self-assembled Peptides for Antimicrobial Therapy

As constructed units, natural amino acids are 20 types to compose peptides and proteins. All the amino acids are similar in structures but different with side groups. Hence, the amino acids can be divided into polar/nonpolar, positively/negatively charged, hydrophobic/hydrophilic, aliphatic/aromatic classes based on the side chains. Besides, except glycine, extra 19 amino acids are chiral molecules. Such variation endows the arrangement of amino acids into huge numbers of peptides or proteins with different biofunctions. Of course, the diversified arrangements are mainly divided into primary structure (e.g., sequences, length, and cyclizing), secondary structures (e.g., α-helix, β-sheet, turns, and coiled coils), and tertiary structures (e.g., geometric shape). As the building blocks for self-assembled nanomaterials, the driving force during the self-assembled process of peptides is weak interactions including hydrogen bonds, hydrophobic affinity, π−π stacking, electrostatic interaction, and van der Waals interactions [20], which are sensitive to pH, temperature, ions, ionic strength, light, etc. As a self-assembled precursor, peptides can be assembled into various kinds of supramolecular nanosized architectures based on design, such as nanotubes [21–23], nanofibers [24–26], nanovesicles [4, 27], nanosheet [28], or hydrogel nanonetworks [29–31]. With modification, peptide hybrid also can assemble into multilayer structure [32, 33] on surface with antibacterial or antifouling functions. Commonly, the antimicrobial properties of these peptide-based nanoassemblies are attributed to rich of cationic for bacterial membrane interference [34]. Besides, the specific structure inspired from nature also contributed to bacterial membrane disruption, such as cell-penetrating peptides [35, 36], lipopeptides [37, 38]. The typical peptides, antimicrobial peptides (AMPs), originally produced by invertebrate, plant, and animal species are amphipathicity, cationic charged, and suitable size for inserting into membrane by "barrel stave," "carpet," or "toroidal pore" mechanisms companied with metabolic inhibition [39]. Based on these speculated mechanisms, the nanoassemblies from peptides or modified peptides can be architected and built with enhanced antimicrobial property.

## 3.2.1 Design and Antimicrobial Mechanism of the Peptide-Based Nanomaterials

In this section, we summarized the design concept and antimicrobial mechanism of the peptide-based nanomaterials. Here, the main functional domains of the building blocks are antimicrobial peptides (AMPs), which can be divided into α-helical, β-sheet, and extended structures [40]. These AMPs are an integral part of the innate immune system and play a role of host defense from invading pathogenic bacteria. Because of the better presentation in the drug resistance, AMPs are now been seen as potential alternatives for antibiotics [41]. Although AMPs vary in sequences, length, and secondary structure, all the peptides adopted a distinct membrane-bound amphipathic conformation. However, the disadvantages of AMPs are short half-life, hemolysis, high expensive, etc; people attempt to overcome these shortages; one of the strategies is nanosized construction of these AMPs [42]. Therein, the self-assembled AMPs perform with nanotubes, nanospheres, nanofibrils, nanotapes, nanosheets, hydrogel nanonetworks, and other ordered structures extremely stabled AMPs and enhanced membrane interactions [38, 43].

To further understand the antimicrobial mechanisms, we initially introduced the action modes of AMPs. Insight of the physical features of AMPs, cationic charge and significant proportion of hydrophobic residues are the general two features. The positively charged property promotes selectivity of the negatively charged microbial cytoplasmic membranes over zwitterionic mammalian membranes, and the hydrophobic residues facilitate interactions with fatty acyl chains [44]. The action modes for disruption of the integrity of the microbial cytoplasmic membrane are toroidal pore-forming, carpet model, barrel stave, electroporation, non-lytic membrane depolarization, anion carrier, oxidized lipid targeting, non-bilayer intermediate, charged lipid clustering, membrane thinning/thickening, disordered toroidal pore-forming [37, 45] (Fig. 3.1).

Besides membrane disruption mechanism, some other antimicrobial mechanisms have now been known similar to antibiotic, such as target key cellular process (DNA and protein synthesis, protein folding, enzymatic activity, and cell wall synthesis) [39]. Interestingly, some findings reveal that AMPs in nature display multiple modes of action to increase the antimicrobial efficiency and evade resistance [46].

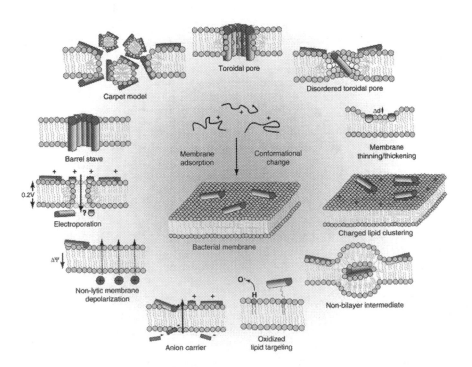

**Fig. 3.1** Action modes of the AMPs during disrupting of microbial cytoplasm membrane

Through the nanoscaled peptide-based self-assembly, the concentrated cationic locally and well-ordered secondary structures arrangement significantly induced "proton sponge" and "multivalence" effect in source of the nanostructures [47–49]. Consequently, the multiple antimicrobial mechanisms all contributed to the enhancement of the antimicrobial property. Of course, the nanoassembly also can be used as stimuli-responsive AMPs carriers to increase the half-life, molecular release specificity, and usage efficiency of AMPs [50–53]. Right now, some mechanism of nanostructures is unknown and still needs exploring, but the excellent antimicrobial property and lower drug resistance of the nanomaterials support us an efficient approach for antimicrobial agents design using nanotechnology.

## 3.2.2  Supramolecular Peptide Amphiphilic Self-assembles as Antimicrobial Agents

The most important design element of peptide amphiphiles (PAs) is amphiphilic nature of the molecules, similar with the lipid or surfactant. The amphiphilicity of PAs results from hydrophobic tails (e.g., alkyl, Fmoc residues, cholesterol) and hydrophilic head (e.g., Arg-rich peptide sequence, TAT) [43, 54, 55], or amphiphilic β-hairpin structure [31, 45]. The second design element is the short sequence directly adjacent to the hydrophobic segment, which is usually composed of hydrophobic amino acids with strong intermolecular hydrogen bonding interactions. Depending on the basic design, the PAs enable to form self-assembled nanostructures from 1D to 2D or 3D [56]. The third design element is additional positively charged amino acids to increase the cationic-rich property. These peptides can be triggered to self-assembly with modulation of pH and ionic strength to realize kinetic growth [40, 57]. The final element for molecular design is the incorporation of different functional domains (e.g., epitopes, enzymatic cleavage sites, disulfide bond) that can be tailored under certain conditions without changing geometry [32, 58] of nanomaterials or triggered to form nanostructures [36, 59].

The driving forces for PAs growth arise from three major non-covalence interactions: hydrophobic interactions, hydrogen bonding, and electrostatic interactions [60]. The assembly behavior in water is normally in a kinetic way from tapered shape of a cluster to sheet-like structure of a group PAs and finally linked together through an intermolecular hydrogen bonding, which determines the interfacial curvature, resulting nanofibrous, nanotubes, nanosheets, nanoparticles, etc. Some of them, such as nanoparticles or nanofibrous, will step-by-step from nanosized structures into 3D hydrogel networks eventually [4].

As for antimicrobial applications, the PAs are normally designed with highly positively charged building blocks to enhance the membrane interference [61]. Other than the monomer of AMPs, the assembled PAs called multidomain peptides (MDPs) with characteristics of cationic and amphiphilic property can be formulated into soluble supramolecular nanostructures [62] (Fig. 3.2).

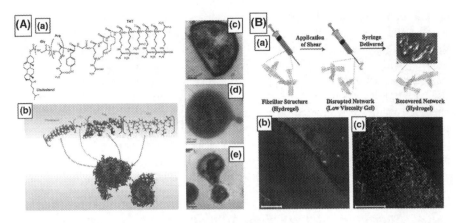

**Fig. 3.2** Assembled PAs with different nanostructures. **A** Self-assembled cationic peptide nanoparticles. Chemical structure of the designed peptide amphiphiles (**a**). The micelle formation simulated with molecular modeling (**b**). TEM images of different bacterial strains: *S. aureus* (**c**), *E. faecalis* (**d**), and *C. neofomans* (**e**) after treatment. **B** β-Hairpin peptide hydrogel. The folding and self-assembly process (**a**) and antibacterial activity with live/dead stain kit (**b**, **c**)

A typical example is a general formula of $K_x(QL)_yK_z$ [63]. The superstructure of such a sequence of peptides is stabled by intermolecular hydrogen bonding and hydrophobic interactions between the (QL) repeating units and electrostatic repulsion between K residues. This supramolecular PAs assembly strategy provides us a platform for antimicrobial agent design with balance of structure, stability, cytotoxicity, and antimicrobial activity. Another example is supramolecular assembly of host defense peptides (HDPs) by modifying an alkyl tail endowed with amphipathicity [38] or utilized natural self-assembled candidates [64]. The well-ordered arrangement of the HDPs with nanofibrous superstructure significantly amplifies the negligible antimicrobial activity of HDPs monomers to pronounced effect. Of course, the amphipathicity of HDPs can also be adjusted by aromatic tag [65] or cholesterol [43, 54] to induce nanovesicle/nanoparticle superstructures with enhanced membrane destabilization and disruption.

The approach of gelatinization of cationic peptides is considered to be effective way for enhancement of antimicrobial activity, such as β-hairpin hydrogel [31, 47] and Fmoc-peptide cationic hydrogel [40]. The mechanism of actions of such hydrogel is debatable; the membrane permeation or disruption of this 3D polycationic network seems to be the common theme. Otherwise, other strategy also supported the antimicrobial activity is in source of the soluble peptides diffused form hydrogel other than its surface interactions [31].

Although the mechanisms for the assemblies above still need better understanding, unquestioned fact is that self-assembly approach significantly strengthens the antimicrobial activity, which seems to be promising direct for antimicrobial agent design.

### 3.2.3 Multilayer Self-assembly of Hybrid Peptides for Surface Antimicrobial Defense

Insight of the nature, multilayered regulation of protein families including defensins, cathelicidins, and C-type lectins constructed the intestinal antimicrobial defense [10]. Thus, as effected approaches, layer-by-layer (LBL) and multilayer assembly obtain considerable development for surface modification. As an important field, artificial surface modifications with antifouling, anti-biofilm, and antimicrobial property through multilayer regulation are significant in microbial resistance.

A most usual way is incorporation of AMPs or HDPs into amphiphilic polymer multilayers. One antimicrobial action is controllable release of peptides with continuing antimicrobial activity [50, 51]. The others are adhering and capturing microbial on the polyelectrolyte multilayers and following high-efficient bacterial killing through surface covalence or locally released peptides [33, 66]. In addition, other than direct microbial killing, a gentle approach is to protect the modified surface from microbial adhesion and anti-biofilm formation [32, 67]. Commonly, peptides are covalent coupling onto polymers and the hybrid polymer–peptides supramolecular assembly with well-organized superstructure on the surfaces.

## 3.3 Polymer Assemblies with Antimicrobial Activity

Polymer chemistry included a hung polymer family with wide range of application in biological fields. Focusing on microorganisms, polymer exhibits unique characteristics than traditional small molecules as antimicrobial agents or bacterial detection agents. With antimicrobial activity, antimicrobial polymers (also named as polymeric biocides) are an area with growing attention. In the health care and hygienic field, biocides polymers are incorporated with other bactericide and can be used for contact disinfectants in many biomedical applications [68].

Here, we just summarized polymers which can self-assemble into nanostructure with antimicrobial activity. Therein, biomimic polymers [34, 69], copolymers with amphiphilicity [70, 71], and hyperbranched polymer [72, 73] are mostly involved. The candidates for self-assembled antimicrobial polymers include natural source (e.g., chitosan [74, 75], peptidopolysaccharides [76], hyaluronic acid [77]) and artificial made (e.g., polyethylenimine [78], poly(amidoamine) [72], polycarbonate [70, 79], polymethacrylates [69, 80]). With delicate design, various supramolecular nanostructures from zero-dimension (0D) to three-dimension (3D) can be constructed through the solution or interfacial self-assembly of amphiphilic polymers, such as micelles, fibers, tubes, vesicles, rod, membranes, and gels. The antimicrobial mechanisms are different according to charge, shape, size, etc., factors, specifically for instance: "anion sponge" membrane suctioning ability through cationic hydrogel [34], distinctive assembled nanosphere and nanorod with efficient electrostatic interaction with microbial cell walls and membranes [42], nanosized micelles with highly dynamic biodegradable capability enable to lysing bacterial membrane [81]. In summary, polymer itself or hybrid with antimicrobial agents would produce a kaleidoscope effect during self-assembly with combined multiple antimicrobial interactions, which show a potential increase in antimicrobial activity, less microbial resistance, excellent biocompatibility and easy to quantity production.

### 3.3.1 Self-assembled Polymers as Antibacterial Agents

The widely accepted theory for antimicrobial polymers was highly positively charged with bacterial membrane disruption. Additionally, the selectivity toward cationic polymers to microbial cells is also a consideration based on different lipid composition and surface components (Fig. 3.3). As seen, the cytoplasmic membranes of mammalian cell are asymmetric and show neutral (zwitterionic) lipid in the outer leaflet of the membranes. In contrast, bacterial and fungal cell membranes exhibit negatively charged component and cover different structural cell walls. Thus, the synthesized polymers and collected natural polymers are always with cationic functionalities [82]. To enhance the antimicrobial activity, AMPs, antibiotics, etc., can emerge simultaneously with polymers through encapslulating [72, 73].

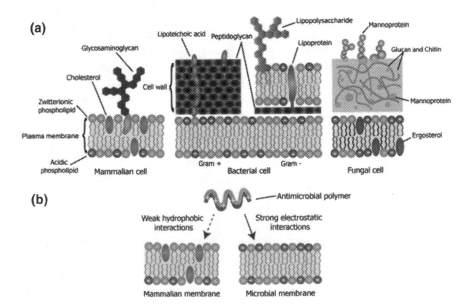

**Fig. 3.3** Selectivity of polyelectrolytes to bacteria over mammalian cells. **a** Illustration of cell membranes of mammalian cells, bacterial cells, and fungal cells. **b** Selective interactions of cationic polymers to different membranes

Other considerations of using polymeric carriers are reducing the toxicity of drugs to human cells and enhancing bacterial targeting. Therefore, the strategies are emerged, such as the nanosized polymer assembly and copolymerization to modify, etc. For development, both antimicrobial and biocompatible polymers are essential for widespread systemic evaluation.

### 3.3.1.1  Structural Design of Cationic Polymers

Unlike small molecule antimicrobial agents, polymers exhibit a typical architecture divided into homopolymers, copolymers with random and block forms, telechelic polymers, zwitterionic polymers, branched polymers, and networked polymers [83] (Fig. 3.4). For structural analysis, it is implied that facial amphiphilicity and hydrophobic content are two significant factors strongly correlated with antimicrobial activity and selectivity. Meantime, the amphiphilicity refers to as hydrophobic/hydrophilic balance, which is influenced by various chemical and physical features of the molecules. Therein, facial amphiphilic secondary structure of the polymer induces helix structure leading to an efficient antimicrobial property similar with some AMPs. Otherwise, higher hydrophilic or dense cationic of the polymers is preferred for binding to cells, but hydrophobic domains are essential for breakthrough cell walls and inserting into bilayer core. Consequently, structural design of antimicrobial polymer is necessarily required.

Firstly, most commonly reported polymers are random copolymers with unspecified sequences. In their structures, charged and hydrophobic moieties are randomly isolated along the polymer backbone. For antimicrobial, random polymer usually includes functionalized poly(methacrylate)s, poly (vinyl ether)s, poly(phenylene vinylene)s, poly(carbonates)s, poly(norbornene)s, poly(methacrylamide)s,

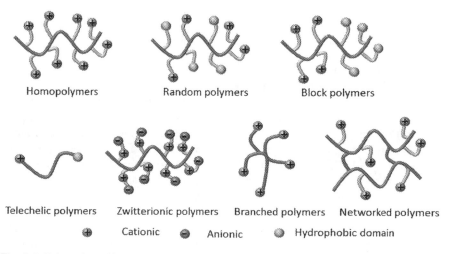

Homopolymers        Random polymers        Block polymers

Telechelic polymers    Zwitterionic polymers    Branched polymers    Networked polymers

⊕ Cationic    ● Anionic    ● Hydrophobic domain

**Fig. 3.4** Polymeric architectures with antimicrobial activity

poly($\beta$-lactam)s [83–85]. Secondly, for homopolymers, there are two typical types: (i) brush-like architecture containing facially amphiphilic repeat units on the backbone; (ii) "same-centered" architecture having the repeat units of hydrophobic moieties accompanied with charged moieties. For example, poly(ethylene imine) with functional carbonates was an amphiphilic homopolymer candidate with antibacterial activity against Gram-positive bacteria [78]. Thirdly, block polymers exhibit strong partitioning between hydrophobic and hydrophilic region along the polymer backbone with a controllable hydrophobic/hydrophilic balance. Usually, these kinds of polymers should self-assembly into well-ordered nanostructures with critical micelle concentration (CMC). For instance, Du' group developed amphiphilic triblock copolymers, which self-assembled into water-dispersible and biodegradable polymer micelles with good antibacterial activities [86]. Fourthly, telechelic polymers are composed of non-biocidal repeating units and reactive end groups, such as block/graft copolymers, star, hyperbranched, or dendritic polymers. The advanced architectures of these polymers are able to built based on these units and groups [87, 88]. Poly(oxazoline)s can be prepared with biocidal end groups and an NHs function at the starting end as a telechelic antimicrobial polymers [89]. Fifthly, zwitterionic polymers are novel charged polymeric antimicrobials bearing an equal number of anionic and cationic groups. Yang and coworkers designed and synthesized well-defined pegylated polymers with tertiary amines for antimicrobial applications [90]. Sixthly, branched polymer includes dendrimer, hyperbranched, brush, or star structures, which is benefit for adsorption and binding to microbial membrane. For instance, branched poly(N-isopropylacrylamide) modified with vancomycin induced a coil-to-globule transition can efficient binding to Gram-positive bacteria with a cell wall inhibition property [73]. Seventhly, networked polymers show a 3D network hydrogel structure, such as peptidopolysaccharides hydrogel, and appear as a broad-spectrum antimicrobial activity attributed to the membrane suctioning ability [76].

### 3.3.1.2 Nanosized Self-assembly Architectures

Anyhow polymer architectures, with appropriate hydrophobic and hydrophilic balance induced amphiphilicity, polymer chains can self-assemble into advanced nanostructures. Some studies indicated that self-assembly of polycations might bring more enhanced activity due to immensely localized mass and positive charge. Lipid similar micelles or liposomes exhibited an easier membrane fusion capability [91]. Besides, complex nanoparticle design endows self-assembled polymers extended properties, such as stimuli-responsiveness, transformers ability, biomimic property. Of course, ingenious design of nanostructure can also enhance biocompatibility. With layer-by-layer microcapsules design based on chitosan, substitution and hyaluronic acid showed a synergistic effect for high-efficient bacterial killing [92]. Core–shell supramolecular assembly with gelatin and red cell membranes realized a biomimic property reducing immune clearance and a "on-demand" bacterial infection antibiotic delivery [93]. A dynamic self-assembled micelles from polycarbonate polymers acquired a desirable microbial membrane facilitating membrane disruption [81]. An amphiphilic poly(amidoamine) dendrimers are enabled to form nanosized particles with antibacterial activity [72]. Enzyme-responsive hyaluronic acid nanocapsules containing polyhexanide presented pathogenic bacterial response biocide release for antibacterial therapy [94]. A fluorescent conjugated oligomer–DNA hybrid hydrogels are represented to antimicrobial regulation property by bacterial DNase hydrolysis [95].

### 3.3.2   Polymer Self-assemblies Coating As Infection-Resistant Surfaces

The antimicrobial activity is important for polymer design; however, as surface protection from microbe the antiadhesion and anti-biofilm are higher demanded [10, 96]. This is because that bacteria are easy to adhere to surfaces of materials and eventually formed bacterial biofilm for their protection. In this situation, most efficient biocides lost their chance to contact with bacterial cells. In this field, polymers are superior due to the easy synthesis, high-efficient inactivation, convenient to modify on surfaces, etc. Here, layer-by-layer assembly and hydrogel formation through polymers are discussed. In addition, smart self-assembly cases are also included stimuli-responsiveness.

Inspired by mussel, antimicrobial multilayer are assembled on poly(tetrafluoroethylene) with catechol-functionalized poly(ethylenimine)and hyaluronic acid in a well-controlled manner [77]. Through an electrostatic LBL self-assembly, quaternized chitosan and soy protein isolate were alternately assembled with a nanofibrous manner, which endowed an inhibition of bacteria [75]. Porous organic frameworks (POFs) were developed to form a durable antibacterial polymer coating through quaternary pyridinium type porous aromatic frameworks (PAFs) [97]. PEG and polycarbonate diblock copolymers and polydopamine were co-coated onto silicone surface with multilayer constructions enabled to prevent biofilm formation [70]. The polycationic hydrogel based on dimethyldecylammonium chitosan exhibits an "anion sponge" interaction on surfaces by attraction of sections of anionic microbial membrane into the internal nanopores of the hydrogel [34, 76] (Fig. 3.5). Polycarbonate and poly(ethylene glycol) in situ assembled hydrogel surface coating vis Michael addition showed a broad antimicrobial and antifouling property [79]. With polycation-bearing catechol and poly(methacrylamide)-bearing quinone inducing nanogels self-assembly, the surface coating was obtained by nanogels arrangement with anti-biofilm and antiadhesion ability [98]. Through a host–guest supramolecular self-assembly of cationic poly(phenylene vinylene) derivative, the antibacterial activity can be controlled on demand during surface modification [99].

Interestingly, the surface modification sometime needs an adaptive ability, which can be well designed by polymer assemblies. For instance, thermo-sensitive poly

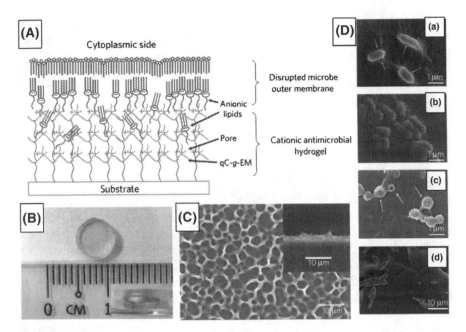

**Fig. 3.5** Surface coating with polycationic hydrogel with microbe membrane suctioning ability. **A** Schematic diagram of the "anion sponge" model. **B** Photograph of the hydrogel-coated substrate. **C** The morphology of the coating by SEM images. **D** Morphology of various pathogens: *P.aeruginosa* (**a**), *E. coli* (**b**), *S. aureus* (**c**), and *F. solani* (**d**), in contact with the hydrogel coating

(N-isopropylacrylamide) assembled modified polymer brushes were endowed with a temperature-response switch capability from bactericidal to cell-repellent [100]. For tissue engineering application, the switch form antifouling to cell adhesion can also be realized by enzymatic response polymer self-assembly [32].

## 3.4 Hybrid Metal/Inorganic Nanomaterials as Bactericide

The metal/inorganic nanomaterials with antibacterial activity usually include $Ag^+$/AgNPs and $TiO_2$ as antibacterial agents hybridizing with organic materials (e.g., polymers, antibiotics, peptides, inorganic salt) and organic antibacterial agents (e.g., antibacterial polymers, antibiotics, antimicrobial peptides, photodynamic/ photothermal agents) hybridizing with inorganic nanomaterials (e.g., silica/silicon nanoparticles, carbon nanotubes, graphene oxide, ferric oxide, gold and copper nanoparticles). Besides, the famous ion: $Ag^+$ and oxidation: $TiO_2$, the investigation reveals that some cationic such as $Ca^{2+}$, $N^+$, $F^+$ and oxidation such as alumina, Ag/Sn/Zn/Pt mixing with $Ar^+$ on polished titanium also exhibit efficient antibacterial effect [101].

Insight of the antimicrobial mechanism of Ag nanoparticles, people find that oxidized silver nanoparticles have antibacterial activities but zero-valent nanoparticles do not, which implies that the antibacterial activities of AgNPs mainly depend on the levels of chemisorbed $Ag^+$ on the particle's surface [102]. Thus, for nanoparticles, AgNPs antibacterial ability exhibits a size dependence behavior. That is, smaller particles size has higher activities on the basis of equivalent silver mass content [103]. Of course, the particles size of AgNPs not only contributes to antibacterial activity but also effects on biosafety (e.g., cytotoxicity, inflammation, developmental toxicity, and genotoxicity) [104]. Besides, good antibacterial activity of AgNPs is dependent on optimally displayed oxidized surfaces, which should be present in well-dispersed medium. Other than size, the shape of AgNPs additional affects their antibacterial activities. For instance, truncated triangular AgNPs with a {111}c lattice plant display a strong biocidal action compared with spherical and rod-shaped NPs [105]. Although a fact has been obtained that $Ag^+$ shows strong inhibitory, bactericidal effect, and antibacterial activity with broad spectrum, the detailed mechanistic study was rare since Kim group supplied that antibacterial mechanism of $Ag^+$ was in source of DNA lost its replication ability and protein became inactivated after $Ag^+$ treatment [106].

The mechanism of the other famous candidate, $TiO_2$, is widely recognized with a photocatalytic reaction, which is straightforward. Upon absorption of photons, electrons of $TiO_2$ are excited from the valence band to the conduction band, creating electron–hole pairs. These charge carriers' photodecompose the surface-adsorbed chemicals, in this process, some radicals or intermediate species (e.g., $\cdot OH$, $O^{2-}$, $H_2O_2$ or $O_2$) play important roles for antibacterial activity [12]. When interacting with bacterial cells, these radicals or intermediate species promote peroxidation of the polyunsaturated phospholipid component of the lipid membrane inducing major disorder in bacterial cell membrane [107]. For photocatalysis improvement, $TiO_2$ nanomaterials can be doped with Ag, Sn, Zn, Pt, etc., or surface implanted with $Ca^{2+}$, $N^+$, $F^+$, etc., [101, 108].

For synergistic interactions, metal, inorganic, and organic components would be hybrid for an improvement of the antibacterial activity or a decrease of the cytotoxic to mammalian cells. In addition, some excite applications of these hybrid are as the bacterial detection sensitizers assistant with well-employed devices (e.g., microfluidic chips, micro-NMR) [109, 110] and technologies (surface-enhanced Raman scattering (SERS), electrochemical system, surface plasmon resonance (SPR), Raman spectroscopy, PCR quantitative identification, mass spectrometry, etc.) [111, 112].

### 3.4.1 Ag Nanoparticles Dispersive into Self-assembled Nanostructures for Synergistic Bacterial Inactivation

A wide application for Ag nanoparticles is concentrated in nanostructures decorated with AgNPs for in vitro or in vivo surface protection. In this fields, nanofibers

[113, 114], hydrogel [115, 116], layer-by-layer self-assembly [117, 118], or nanoparticles arrays [118, 119] can be used as scaffolds to directly disperse Ag nanoparticles or mineralize to form monodispersive AgNPs. With a synergistic interaction, highly dispersed AgNPs display a long-term antimicrobial activity inside nanoassembled structures. Although the biosafety of the AgNPs in vivo application is still in dispute, these self-assembled AgNPs hybrid nanomaterials display significant roles in biomedical surface protection both for medical instruments and tissue engineering (Fig. 3.6).

There are some typical examples reported in recent 5 years. The peptide nanofibers as scaffold have been successfully applied in tissue engineering; hence, nucleation/growth of well-ordered AgNPs into peptide-based nanofibers exhibited excellent antibacterial properties [113, 114]. Hydrogels always seemed as good candidates for surface modification. The self-assembled nanocomposites consisted of polymeric [115] or amino acids [116]-based hydrogel and dispersive AgNPs enable to be well spread on surfaces to form an antifouling and antibacterial surface coating. Cooperated with cationic polymers, which is superior in selective adhesion

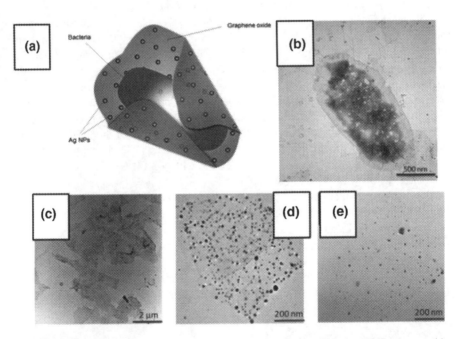

**Fig. 3.6** Hybrid nanostructures decorated with silver nanoparticles. **A** AgNPs on peptide nanofibers with morphology (**a**), antibacterial activity to *E. coli* (**b**, **c**), and structure illustration (**d**). **B** DNA-directed AgNPs on graphene oxide (Ag@dsDNA@GO) with structure illustration (**a**), bacterial killing ability (**b**), morphology of bare GO (**c**), Ag@dsDNA@GO composite with diameter of Ag of $\sim 18 \pm 3$ nm (**d**), and $\sim 5 \pm 1.8$ nm (**e**). **C** Silver-decorated shell cross-linked knedel-like polymer nanoparticles with three silver-loaded variants. The structure illustration of the nanoparticles and antimicrobial activity in mouse model (**a**) and TEM images of particles morphology (**b**)

of bacteria cells, the cationic polymer hybrid AgNPs with LBL or multilayers nanostructures on surface appeared as a highly efficient bacterial killing strategy because of the synergistic effect [117–119]. For water-dispersible applications, the AgNPs decorated polymer vesicles/micelles [120], graphene oxide [121], carbon nanotubes [122], or silicon nanowires [123] were well considered as the hybrid carrier to immobile and disperse AgNPs.

### 3.4.2 Others Hybrid Nanoassembles as Antibacterial Agents

One of the optimized strategies for AgNPs was hybrid with other metal nanostructures, such as silver-coated copper [124] and magnetic nanoparticles [125, 126], silver heterodimers [127], and Ag@Co-nanocomposites [128]. Of course, besides silver, other metal/inorganic hybrid nanomaterials also display significant role as antibacterial agents. For instance, $TiO_2$ hybrid arrays [129], reactive oxygen species located in mesoporous inorganic nanomaterials [130], core–shell assembled complex with polymers [131–133] /proteins [134, 135] /peptides [136] /amino acids [137] are all demonstrated to be the efficient antibacterial strategies. Especially, Au nanoparticles hybrids exhibit typical photothermal capability for thermal bacterial killing [138, 139]. Here, we exampled some typical cases. One is self-assembly of polyphenol-based core/shell nanostructures, which strongly inhibited E. coli growth while exhibited no obvious normal cell toxicity [132]. A visible-light-controlled drug release TiNTs array was constructed with decorated Au nanoparticles, hydrophobic octadecylphosphonic acid (ODPA), and AMPs, by which the antibacterial activity can be active by visible-light irradiation [129]. By high loading efficacy of mesoporous organosilica of reactive oxygen species and nitric oxide, these sunlight-triggered nanoparticles showed synergistically antibacterial activity based on sliver-free ability [130]. An interesting example was that fabricating shape-selective thermal killing nanoassemblies via Au nanoparticles, silica, and colloid antibodies hybrids [139].

With complex supramolecular self-assembly, many hybrid nanomaterials will be assembled synergistically with enhanced antibacterial efficiency. In addition, well-controlled arrangement of each antibacterial element has been demonstrated to contribute more to antibacterial activity. With dramatic development of various antibacterial nanomaterials, we believe that hybrid nanomaterials could occupy an important place in future.

## 3.5 Photodynamic Antibacterial Agent Assembled Nanomaterials

Normally, photodynamic therapies (PDT) are dependent on the photosensitizers, such as porphyrin derivatives, BODIPY derivatives, cationic phenothiazine dyes, and anthraquinone vat dyes. The mechanism of these photosensitizers is easy to understand, which can be light-activated to reduce reactive oxygen species (ROS) in aqueous phase resulting in a chemical inactivation of bacteria [140]. As known, the lift time of ROS was short and the interaction distance has exact demands [141]. And the aggregation of these molecules may dramatically decrease their photodynamic capability [142]. For high effective photodynamic therapy, self-assemble of the molecular derivatives or decorated those into assembled structures are effective strategies. Concretely listed, there are modified molecules endowing amphiphilicity [7, 143–145], supramolecular assembly [146], and polymer conjugation for self-assembly [147–150]. Here, we exampled typical cases for better understanding. Through a host–guest interaction to form supramolecular assemblies of porphyrin derivatives (TPOR, with four positive charges), the supramolecular photosensitizers acquired greatly improved antibacterial efficiency [146]. When encapsulated Cp6 into charge-conversion polymeric nanoparticles, the complex nanosystem showed acidic-induced charge conversion in bacterial infection site for efficient accumulation and then photodynamic inactivation for therapy [150].

## 3.6 Antibiotic Assembly with Enhanced Antibacterial Efficiency

The wide application of antibiotic is small molecules. However, the antibiotic resistance and superbug appearance are now becoming intensive. The usage of antibiotic is facing severe challenges. To avoid the over usage of antibiotics, the traditional drug delivery strategy by assembled nanomaterial supply is a solution. The hydrophilicity of the antibiotic other than usually displayed hydrophobic drugs demands a higher need for control release. Thus, water-soluble proteins or polymers with multilayers self-assembly or stimuli-responsive release property can be considered, such as gelatin core–shell nanoparticles with bacterial toxin responsive release capability [93], multilayer nanogel [151, 152] or gelatin sponges [153], and dendrimer-based multivalent platform [154]. More deserved attention strategy is that via molecular modification, antibiotic could self-assembly into nanostructure with enhanced antibacterial efficiency, which attributes to structure effect and multivalence effect [14, 155, 156]. The first antibiotic gelator reported was vancomycin–pyrene from Xu' group [14]. They speculated that polymer-like assembly structures of the hydrogel enhanced the local concentration on the bacterial cell surface, which led to a lower molecule usage comparing with vancomycin against bacteria. The amphiphilic tobramycins self-assemblies not only maintain their antibacterial activity but also can boost the innate immune response [157].

## 3.7  Self-assembled Nanomaterials for Bacterial Detection

Bacterial detection approaches in vitro are involving directly bacterial cell detection [158] and bacterial relative enzymes, proteins, lipopolysaccharides, DNA, RNA, etc. and molecules detection [95, 159–161]. Extended to in vivo bacterial infection diagnosis, the biomarkers not only involved in bacterial themselves [15, 162, 163] and bacterial infectious microenvironments [142, 164], but the host-specific immunity response-related molecules, such as interleukin [165], caspase-1 [36]. The optimized specificity of all the methods in vitro is bacterial DNA/RNA detection and identification, by which the strains of different bacteria cell can be defined [166]. In addition, for convenient application in food safety and clinical screening, the colorimetric/fluorescence visualized detection methods are more popular [19, 167]. For in vivo applications, the pathogen detection methods are limited, mostly concentrated in bioimaging. Therein, the technologies for bacterial detection are divided into two parts: in vitro and in vivo applications. In vitro electrochemical/optical spectroscopy techniques are normally used for detection [168–172]; magnetic technique is superior for trapping and separation [162, 173–175]; PCR technique is preferred for identification [176, 177]. However, for living body, the in vivo imaging technologies are welcomed, such as in vivo imaging system (including bioluminescence and fluorescence imaging) [178, 179], photoacoustic tomography (PAT) [36, 142], and positron emission computed tomography (PET/CT) [180, 181]. Thus, to enhance sensitivity and specificity for bacterial detection based on these technologies, self-assembled nanomaterials enable to be designed as biosensors [160, 169, 170], bacterial tracers [15], imaging probes [179, 182], etc. In this section, we will summarize different kinds of self-assembled nanomaterials for bacterial detection in vitro and in vivo to overlook the materials' development in this field.

### 3.7.1  Polymer-Based Assemblies to Detect Bacteria

Polymer science in bacterial detection presents a promising development recently. Besides for sensitive bacterial detection technology development [110, 111, 169], bacterial sensors with nanomaterials are also important for contributing to enhance the sensitivity, specificity, and identification [109, 179, 183, 184]. In different with small molecular tracers, which are specific targeting to bacterial membrane (includes membrane-bound enzymes and membrane-bound receptors), intracellular enzymes, DNA synthesis, and translational machinery [15], polymeric self-assembled nanomaterials are endowed with multiple interactions to increase specificity, responsive release signal molecules to amplify the imaging signals, and bacterial infectious microenvironment specificity to identify bacterial species. The triggered factors involve bacterial cytotoxins [185], negative-charged bacterial membranes [17, 169, 186], bacterial lipopolysaccharides (LPS) [187], extracellular enzymes [160, 188, 189], and pH [190]. For example, agarose hydrogel network self-assembled fluorescent dye vesicles [185] and red cell membrane-coated gelatin self-assembled nanoparticles [93] can be triggered to release the carried contents by bacterial cytotoxin, especially pore-forming toxin secreted by some strains with hemolysis during infections. The development of PDA/phospholipids/lysine vesicles with different pentacosadiynoic acid (PCDA) contents showed a different specificity to bacterial strains based on electrostatic interaction with lysine and bacterial membranes for detection [17]. Using peptide functionalized polydiacetylene liposome with fluorescence probe inserted, the quenched fluorescence can be selectively turned on after interaction with LPS or bacterial surfaces [159] (Fig. 3.7). For bacteria sensing, polymersomes assembled by hyaluronic acid-*b*-poly(ε-caprolactone) copolymers enable to hyaluronidase responsive to release the fluorescence cargos with an amplified signal [188]. Dependent on the acidic environment induced by bacterial infection, a sensitive pH-responsive polymeric micelle consisting of poly(ethylene glycol-*b*-trimethylsilyl methacrylate) was built for sensitive and selective bacterial detection [190].

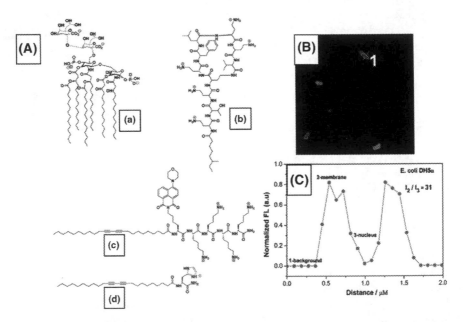

**Fig. 3.7** Polymer-based liposomes as a sensor for bacterial LPS. **A** Structures of lipid (**a**), polymyxin (**b**), and the two peptide amphiphiles (**c**) and (**d**). **B** The confocal luminescence image of the E. coli DH5α after incubating with liposomes. **C** The luminescence intensity profile of the confocal image

### 3.7.2 Metal/Inorganic Hybrid Assemblies as Bacterial Sensor

The hybrid assemblies as bacterial detection biosensor can be divided into two styles—one is surface assemblies to modified sensor devices [171, 191], and the other is complex assemblies dispersed in solution [172, 192]. The surface modification approach for detection is aimed to specific recognization of bacterial cells or DNA, etc. and amplified the readout signals by metal/inorganic nanomaterials, for example, for electrochemical sensor, graphenes, and Au nanoparticles hybrid assembly with bacterial affinity molecules to capture cells with signal amplification [191, 193, 194]. For instance, a supramolecular enzyme–nanoparticle was fabricated, by which the colorimetric readout can be enzyme-amplified [160] (Fig. 3.8). Additionally, Au nanoparticles hybrid hairpin-structured assembly can be triggered to detect DNA sequence with enhanced SERS signals [171]. It is implied that using a monolayer graphene oxide self-assembly sheet, the field-effect transistor sensor devices can be highly sensitive and selective detected bacteria [195]. Besides, the typical example for bacterial sensor in solution was that based on self-assembled DNAzyme labeled DNA probes with SWNT conjugates, the bacteria detection can be visualized readout by colored product in solution [196]. Another case was a "light-up" supramolecular nanoprobe assembled by pyrene-labeled peptide amphiphiles hybrid with silver ions [197].

### 3.7.3 Others Self-assembled Probes

The platforms as probes for bacterial detection are normally dependent on fluorescence or colorimetric signal readouts. Thus, the stimuli-responsive strategies,

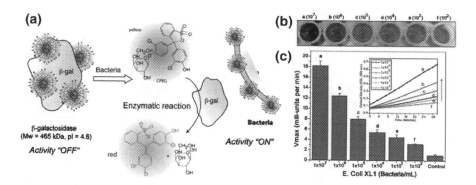

**Fig. 3.8** Au hybrid supramolecular enzyme–nanoparticles for colorimetric bacteria sensing. **a** The illustration of enzyme-amplified sensing of bacteria. **b** The photograph of the colors and **c** kinetic absorbance responses upon addition of different concentration of bacteria cells

such as aggregation-induced emission (AIE) characteristics, FRET effect, light-up
with cleavage of the fluorescence quenchers, structural insert-induced light-up,
deassembly induced turn-on, enzyme-assisted fluorescence signal amplification, are
widely considered [19, 198, 199]. For instance, through target binding, an assembly
of multiple DNA components can be used for specific detection of bacterial proteins
[200]. A supramolecular platform based on self-assembled monolayers (SAMs) had
been implemented in a microfluidic device for anthrax detection [201].
Interestingly, some strategies exhibited a novel concept, which is bacterial
interaction-induced assembly or transformers for detection. For examples, a NBD–
vancomycin conjugates showed bacterial interfacial self-assembly to form
fluorescent nanoparticles for bacterial detection [155].

### 3.7.4   In Vivo Self-assembly Strategy for Bacterial Infection Diagnosis

The concept of in vivo self-assembly is newly presented recently. The essential
difference of the concept to assembly is that the self-assembly is not initially
constructed before introduced to living subjects, but "smart" self-assembled in
certain region in vivo. Why we need in vivo self-assembly? Compared to small
molecules, nanomaterials exhibit excellent behavior in sensitivity and specificity for
detection, while the small molecules is superior in reliability, stability, and biocompatibility. It is obvious that combining both advantages of small molecules and
nanomaterials, in vivo self-assembly strategy represents huge potential in biological
application, especially for bacterial infection imaging (Fig. 3.9). Some studies
indicated that enhanced permeability retention (EPR) effect exists in infection
[164]. However, the nanosized bacterial detection probes in vivo are rarely reported. We believe that the immune cleavage and the poor EPR effect greatly limited
the nanomaterial in vivo. Thus, the effective bacterial infection imaging probes
reported in recent 10 years are all small molecules [15, 178, 179, 202, 203].
Therein, the best detection limit was $10^5$ colony-forming units bacterial infection
in vivo reported by Murthy group at 2011, which was published on Nature
Materials [202]. However, based on our previous study of the in vivo self-assembly,
we found that the in vivo assembly will induce high retention effect named AIR
effect, which not only enhanced molecular retention efficiency but increased
specific selectivity [204]. In this regards, we designed a bacterial-specific response
small molecule: Ppa-PLGVRG-Van, which can firstly high efficient target to bacteria cell walls, then the molecules were tailored by bacterial released gelatinase,
finally the molecular residues are well-ordered to assembly into J-type aggregates
with enhanced photoacoustic signals. This de novo strategy dramatically increased
the sensitivity and specificity for bacterial infection imaging and reduced the
detection limits to $10^3$ colony-forming units, which was two magnificent lower than
reported before [142]. For extension, this strategy can also be used for noninvasive

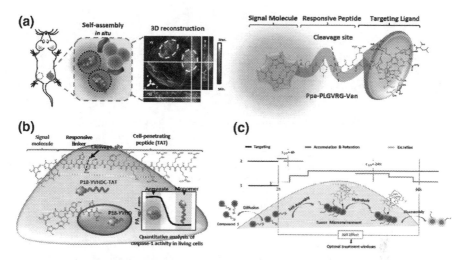

**Fig. 3.9** In vivo self-assembly with enhanced photoacoustic signal for bacterial detection. **a** Highly sensitive and specific bacterial infection imaging based on Ppa–peptide derivatives. This strategy enables both image bacterial infection site with 3D reconstruction and identifies different bacterial infection (such as gelatinase positive/negative strains and sterile myositis). **b** Quantitative analysis of caspase-1 inside living cells as an early bacterial infection alert. **c** The mechanism illustration of aggregation-induced retention (AIR) effect by chlorophyll–peptide-based nanoassemblies

quantitative caspase-1 imaging intracellularly for early bacterial infection alert [36]. As a forehead self-assembly direction, we have faith that in vivo self-assembly certainly will lead the trend of the supramolecular science during the following years.

## 3.8  Conclusion Remarks

Now, we have opened an era of "molecule domestication" to fabricate various structures, functions, and response self-assemblies for bacterial infection diagnosis and therapy. Facing the huge challenges of task risks, self-assembled nanomaterials might explore a promising way. Insight of nature, biological subjects spontaneously regulated their biofunctions through nature tailored molecular assemblies and dis-assemblies to adapt the change of environment for evolution process. Inspired by nature, the stimuli adaptability molecular assembly or even regulated the "smart" assembly units to self-assembly in vivo can be fabricated by integrating the structures and biofunctions together to better biomimic nature. The exciting odyssey begins; all of us need interdisciplinary cooperation and unremitting efforts to promote the self-assembly biomaterials for bacterial infection diagnosis and therapy application.

# References

1. Levy SB, Marshall B (2004) Antibacterial resistance worldwide: causes challenges and responses. Nat Med 10(12 Suppl):S122–S129
2. Costa F, Carvalho IF, Montelaro RC, Gomes P, Martins MCL (2011) Covalent immobilization of antimicrobial peptides (AMPs) onto biomaterial surfaces. Acta Biomater 7(4):1431–1440
3. Zhao X, Pan F, Xu H, Yaseen M, Shan H, Hauser CAE, Zhang S, Lu JR (2010) Molecular self-assembly and applications of designer peptide amphiphiles. Chem Soc Rev 39(9):3480–3498
4. Gazit E (2007) Self-assembled peptide nanostructures: the design of molecular building blocks and their technological utilization. Chem Soc Rev 36(8):1263–1269
5. Moon RJ, Martini A, Nairn J, Simonsen J, Youngblood J (2011) Cellulose nanomaterials review: structure, properties and nanocomposites. Chem Soc Rev 40(7):3941–3994
6. Zhou Y, Huang W, Liu J, Zhu X, Yan D (2010) Self-Assembly of hyperbranched polymers and its biomedical applications. Adv Mater 22(41):4567–4590
7. Liu F, Soh Yan Ni A, Lim Y, Mohanram H, Bhattacharjya S, Xing B (2012) Lipopolysaccharide neutralizing peptide-porphyrin conjugates for effective photoinactivation and intracellular imaging of gram-negative bacteria strains. Bioconjug Chem 23 (8):1639–1647
8. Silva ZS Jr, Bussadori SK, Fernandes KP, Huang YY, Hamblin MR (2015) Animal models for photodynamic therapy (PDT). Biosci Rep 35(6):e00265
9. Sengupta S, Würthner F (2013) Chlorophyll J-Aggregates: from bioinspired dye stacks to nanotubes, liquid crystals, and biosupramolecular electronics. Acc Chem Res 46(11):2498–2512
10. Mukherjee S, Vaishnava S, Hooper LV (2008) Multi-layered regulation of intestinal antimicrobial defense. Cell Mol Life Sci 65(19):3019–3027
11. Zhu X, Jun Loh X (2015) Layer-by-layer assemblies for antibacterial applications. Biomater Sci 3(12):1505–1518
12. Chen X, Mao SS (2007) Titanium dioxide nanomaterials: synthesis, properties, modifications, and applications. Chem Rev 107(7):2891–2959
13. Tarpley RJ (2014) Antibiotics: discontinue low-dose use. Science 343(6167):136–137
14. Xing B, Yu C-W, Chow K-H, Ho P-L, Fu D, Xu B (2002) Hydrophobic interaction and hydrogen bonding cooperatively confer a vancomycin hydrogel: a potential candidate for biomaterials. J Am Chem Soc 124(50):14846–14847
15. Bunschoten A, Welling MM, Termaat MF, Sathekge M, van Leeuwen FW (2013) Development and prospects of dedicated tracers for the molecular imaging of bacterial infections. Bioconjug Chem
16. Wang H, Zhou Y, Jiang X, Sun B, Zhu Y, Wang H, Su Y, He Y (2015) Simultaneous capture, detection, and inactivation of bacteria as enabled by a surface-enhanced raman scattering multifunctional chip. Angew Chem Int Ed 54(17):5132–5136
17. de Oliveira TV, Soares NdFF, Silva DJ, de Andrade NJ, Medeiros EAA, Badaró AT (2013) Development of PDA/Phospholipids/Lysine vesicles to detect pathogenic bacteria. Sens Actuat B Chem 188:385–392
18. Li LL, Ma HL, Qi GB, Zhang D, Yu F, Hu Z, Wang H (2016) Pathological-condition-driven construction of supramolecular nanoassemblies for bacterial infection detection. Adv Mater 28(2):254–262
19. Feng G, Yuan Y, Fang H, Zhang R, Xing B, Zhang G, Zhang D, Liu B (2015) A light-up probe with aggregation-induced emission characteristics (AIE) for selective imaging, naked-eye detection and photodynamic killing of Gram-positive bacteria. Chem Commun 51(62):12490–12493
20. Zhao X, Zhang S (2006) Molecular designer self-assembling peptides. Chem Soc Rev 35(11):1105–1110

21. Xu C, Liu R, Mehta AK, Guerrero-Ferreira RC, Wright ER, Dunin-Horkawicz S, Morris K, Serpell LC, Zuo X, Wall JS, Conticello VP (2013) Rational design of helical nanotubes from self-assembly of coiled-coil lock washers. J Am Chem Soc 135(41):15565–15578
22. Scanlon S, Aggeli A (2008) Self-assembling peptide nanotubes. Nano Today 3(3–4):22–30
23. Rubin DJ, Amini S, Zhou F, Su H, Miserez A, Joshi NS (2015) Structural, nanomechanical, and computational characterization of d, l-Cyclic peptide assemblies. ACS Nano 9(3):3360–3368
24. Cormier AR, Pang X, Zimmerman MI, Zhou H-X, Paravastu AK (2013) Molecular structure of RADA16-I designer self-assembling peptide nanofibers. ACS Nano 7(9):7562–7572
25. Huang C-C, Ravindran S, Yin Z, George A (2014) 3-D self-assembling leucine zipper hydrogel with tunable properties for tissue engineering. Biomaterials 35(20):5316–5326
26. Wu EC, Zhang S, Hauser CAE (2012) Self-assembling peptides as cell-interactive scaffolds. Adv Funct Mater 22(3):456–468
27. Lin BF, Marullo RS, Robb MJ, Krogstad DV, Antoni P, Hawker CJ, Campos LM, Tirrell MV (2011) De novo design of bioactive protein-resembling nanospheres via dendrimer-templated peptide Amphiphile assembly. Nano Lett 11(9):3946–3950
28. Hamley IW, Dehsorkhi A, Castelletto V (2013) Self-assembled arginine-coated peptide nanosheets in water. Chem Commun 49(18):1850–1852
29. Irwansyah I, Li Y-Q, Shi W, Qi D, Leow WR, Tang MBY, Li S, Chen X (2015) Gram-positive antimicrobial activity of amino acid based hydrogels. Adv Mater 27(4): 648–654
30. Veiga AS, Sinthuvanich C, Gaspar D, Franquelim HG, Castanho MARB, Schneider JP (2012) Arginine-rich self-assembling peptides as potent antibacterial gels. Biomaterials 33 (35):8907–8916
31. Salick DA, Kretsinger JK, Pochan DJ, Schneider JP (2007) Inherent antibacterial activity of a peptide-based β-hairpin hydrogel. J Am Chem Soc 129(47):14793–14799
32. Li L-L, Qi G-B, Yu F, Liu S-J, Wang H (2015) An adaptive biointerface from self-assembled functional peptides for tissue engineering. Adv Mater 27(20):3181–3188
33. Guyomard A, Dé E, Jouenne T, Malandain J-J, Muller G, Glinel K (2008) Incorporation of a hydrophobic antibacterial peptide into Amphiphilic polyelectrolyte multilayers: a bioinspired approach to prepare biocidal thin coatings. Adv Funct Mater 18(5):758–765
34. Li P, Poon YF, Li W, Zhu H-Y, Yeap SH, Cao Y, Qi X, Zhou C, Lamrani M, Beuerman RW, Kang E-T, Mu Y, Li CM, Chang MW, Jan Leong SS, Chan-Park MB (2011) A polycationic antimicrobial and biocompatible hydrogel with microbe membrane suctioning ability. Nat Mater 10(2):149–156
35. Liu L, Xu K, Wang H, Jeremy Tan PK, Fan W, Venkatraman SS, Li L, Yang Y-Y (2009) Self-assembled cationic peptide nanoparticles as an efficient antimicrobial agent. Nat Nanotechnol 4(7):457–463
36. Li LL, Zeng Q, Liu WJ, Hu XF, Li Y, Pan J, Wan D, Wang H (2016) Quantitative analysis of caspase-1 activity in living cells through dynamic equilibrium of chlorophyll-based nano-assembly modulated photoacoustic signals. ACS Appl Mater Interfaces 8(28):17936–17943
37. Nguyen LT, Haney EF, Vogel HJ (2011) The expanding scope of antimicrobial peptide structures and their modes of action. Trends Biotechnol 29(9):464–472
38. Ravi J, Bella A, Correia AJV, Lamarre B, Ryadnov MG (2015) Supramolecular amphipathicity for probing antimicrobial propensity of host defence peptides. Phys Chem Chem Phys 17(24):15608–15614
39. Brogden KA (2005) Antimicrobial peptides: pore formers or metabolic inhibitors in bacteria? Nat Rev Micro 3(3):238–250
40. Debnath S, Shome A, Das D, Das PK (2010) Hydrogelation through self-assembly of FMOC-peptide functionalized cationic Amphiphiles: potent antibacterial agent. J Phys Chem B 114(13):4407–4415
41. Li P, Li X, Saravanan R, Li CM, Leong SSJ (2012) Antimicrobial macromolecules: synthesis methods and future applications. RSC Adv 2(10):4031–4044

42. Fukushima K, Tan JP, Korevaar PA, Yang YY, Pitera J, Nelson A, Maune H, Coady DJ, Frommer JE, Engler AC, Huang Y, Xu K, Ji Z, Qiao Y, Fan W, Li L, Wiradharma N, Meijer EW, Hedrick JL (2012) Broad-spectrum antimicrobial supramolecular assemblies with distinctive size and shape. ACS Nano 6(10):9191–9199

43. Liu L, Xu K, Wang H, Jeremy Tan PK, Fan W, Venkatraman SS, Li L, Yang Y-Y (2009) Self-assembled cationic peptide nanoparticles as an efficient antimicrobial agent. Nat Nanotechnol 4(7):457–463

44. Hu J, Chen C, Zhang S, Zhao X, Xu H, Zhao X, Lu JR (2011) Designed antimicrobial and antitumor peptides with high selectivity. Biomacromol 12(11):3839–3843

45. Shenkarev ZO, Balandin SV, Trunov KI, Paramonov AS, Sukhanov SV, Barsukov LI, Arseniev AS, Ovchinnikova TV (2011) Molecular mechanism of action of β-hairpin antimicrobial peptide arenicin: oligomeric structure in dodecylphosphocholine micelles and pore formation in planar lipid bilayers. Biochemistry 50(28):6255–6265

46. Peschel A, Sahl H-G (2006) The co-evolution of host cationic antimicrobial peptides and microbial resistance. Nat Rev Micro 4(7):529–536

47. Salick DA, Pochan DJ, Schneider JP (2009) Design of an injectable β-hairpin peptide hydrogel that kills methicillin-resistant Staphylococcus aureus. Adv Mater 21(41):4120–4123

48. Ong ZY, Gao SJ, Yang YY (2013) Short synthetic β-sheet forming peptide Amphiphiles as broad spectrum antimicrobials with antibiofilm and endotoxin neutralizing capabilities. Adv Funct Mater 23(29):3682–3692

49. Bhatia S, Camacho LC, Haag R (2016) Pathogen inhibition by multivalent ligand architectures. J Am Chem Soc 138(28):8654–8666

50. Shukla A, Fleming KE, Chuang HF, Chau TM, Loose CR, Stephanopoulos GN, Hammond PT (2010) Controlling the release of peptide antimicrobial agents from surfaces. Biomaterials 31(8):2348–2357

51. Cado G, Aslam R, Seon L, Garnier T, Fabre R, Parat A, Chassepot A, Voegel JC, Senger B, Schneider F, Frere Y, Jierry L, Schaaf P, Kerdjoudj H, Metz-Boutigue MH, Boulmedais F (2013) Self-defensive biomaterial coating against bacteria and yeasts: polysaccharide multilayer film with embedded antimicrobial peptide. Adv Funct Mater 23(38):4801–4809

52. Chen L, Liang JF (2013) Peptide fibrils with altered stability, activity and cell selectivity. Biomacromolecules 14(7):2326–2331

53. Li L-l, Wang H (2013) Enzyme-coated mesoporous silica nanoparticles as efficient antibacterial agents in vivo. Adv Healthcare Mater:1351–1360

54. Wang H, Xu K, Liu L, Tan JPK, Chen Y, Li Y, Fan W, Wei Z, Sheng J, Yang Y-Y, Li L (2010) The efficacy of self-assembled cationic antimicrobial peptide nanoparticles against Cryptococcus Neoformans for the treatment of meningitis. Biomaterials 31(10):2874–2881

55. Chen C, Hu J, Zhang S, Zhou P, Zhao X, Xu H, Zhao X, Yaseen M, Lu JR (2012) Molecular mechanisms of antibacterial and antitumor actions of designed surfactant-like peptides. Biomaterials 33(2):592–603

56. Cui H, Webber MJ, Stupp SI (2010) Self-assembly of peptide amphiphiles: From molecules to nanostructures to biomaterials. Pept Sci 94(1):1–18

57. Sikorska E, Dawgul M, Greber K, Iłowska E, Pogorzelska A (1838) Kamysz W (2014) Self-assembly and interactions of short antimicrobial cationic lipopeptides with membrane lipids: ITC, FTIR and molecular dynamics studies. BBA Biomembranes 10:2625–2634

58. Tian X, Sun F, Zhou X-R, Luo S-Z, Chen L (2015) Role of peptide self-assembly in antimicrobial peptides. J Pept Sci 21(7):530–539

59. Li L-L, Ma H-L, Qi G-B, Zhang D, Yu F, Hu Z, Wang H (2015) Pathological-condition-driven construction of supramolecular nanoassemblies for bacterial infection detection. Adv Mater 27(20):3181–3188

60. Ulijn RV, Smith AM (2008) Designing peptide based nanomaterials. Chem Soc Rev 37(4):664–675

61. Torrent M, Valle J, Nogués MV, Boix E, Andreu D (2011) The generation of antimicrobial peptide activity: a trade-off between charge and aggregation? Angew Chem Int Ed 50 (45):10686–10689

62. Chen C, Pan F, Zhang S, Hu J, Cao M, Wang J, Xu H, Zhao X, Lu JR (2010) Antibacterial activities of short designer peptides: a link between propensity for nanostructuring and capacity for membrane destabilization. Biomacromol 11(2):402–411

63. Xu D, Jiang L, Singh A, Dustin D, Yang M, Liu L, Lund R, Sellati TJ, Dong H (2015) Designed supramolecular filamentous peptides: balance of nanostructure, cytotoxicity and antimicrobial activity. Chem Commun 51(7):1289–1292

64. Chairatana P, Nolan EM (2014) Molecular basis for self-assembly of a human host-defense peptide that entraps bacterial pathogens. J Am Chem Soc 136(38):13267–13276

65. Joshi S, Dewangan RP, Yar MS, Rawat DS, Pasha S (2015) N-terminal aromatic tag induced self assembly of tryptophan-arginine rich ultra short sequences and their potent antibacterial activity. RSC Adv 5(84):68610–68620

66. Glinel K, Jonas AM, Jouenne T, Jrm Leprince, Galas L, Huck WTS (2008) Antibacterial and antifouling polymer brushes incorporating antimicrobial peptide. Bioconjug Chem 20(1): 71–77

67. Lim K, Chua RRY, Saravanan R, Basu A, Mishra B, Tambyah PA, Ho B, Leong SSJ (2013) Immobilization studies of an engineered Arginine–Tryptophan-Rich peptide on a silicone surface with antimicrobial and antibiofilm activity. ACS Appl Mater Interfaces 5(13):6412–6422

68. Kenawy E-R, Worley SD, Broughton R (2007) The chemistry and applications of antimicrobial polymers: a state-of-the-art review. Biomacromol 8(5):1359–1384

69. Locock KES, Michl TD, Stevens N, Hayball JD, Vasilev K, Postma A, Griesser HJ, Meagher L, Haeussler M (2014) Antimicrobial polymethacrylates synthesized as mimics of tryptophan-rich cationic peptides. ACS Macro Lett 3(4):319–323

70. Ding X, Yang C, Lim TP, Hsu LY, Engler AC, Hedrick JL, Yang YY (2012) Antibacterial and antifouling catheter coatings using surface grafted PEG-b-cationic polycarbonate diblock copolymers. Biomaterials 33(28):6593–6603

71. Park J, Kim J, Singha K, Han D-K, Park H, Kim WJ (2013) Nitric oxide integrated polyethylenimine-based tri-block copolymer for efficient antibacterial activity. Biomaterials 34(34):8766–8775

72. Lu Y, Slomberg DL, Shah A, Schoenfisch MH (2013) Nitric oxide-releasing Amphiphilic Poly(amidoamine) (PAMAM) dendrimers as antibacterial agents. Biomacromol 14 (10):3589–3598

73. Shepherd J, Sarker P, Swindells K, Douglas I, MacNeil S, Swanson L, Rimmer S (2010) Binding bacteria to highly branched Poly(N-isopropyl acrylamide) Modified with Vancomycin induces the coil-to-globule transition. J Am Chem Soc 132(6):1736–1737

74. Shrestha A, Kishen A (2012) The effect of tissue inhibitors on the antibacterial activity of Chitosan nanoparticles and photodynamic therapy. J Endod 38(9):1275–1278

75. Pan Y, Huang X, Shi X, Zhan Y, Fan G, Pan S, Tian J, Deng H, Du Y (2015) Antimicrobial application of nanofibrous mats self-assembled with quaternized chitosan and soy protein isolate. Carbohydr Polym 133:229–235

76. Li P, Zhou C, Rayatpisheh S, Ye K, Poon YF, Hammond PT, Duan H, Chan-Park MB (2012) Cationic peptidopolysaccharides show excellent broad-spectrum antimicrobial activities and high selectivity. Adv Mater 24(30):4130–4137

77. Lee H, Lee Y, Statz AR, Rho J, Park TG, Messersmith PB (2008) Substrate-independent layer-by-layer assembly by using mussel-adhesive-inspired polymers. Adv Mater 20 (9):1619–1623

78. He Y, Heine E, Keusgen N, Keul H, Möller M (2012) Synthesis and characterization of amphiphilic monodisperse compounds and Poly(ethylene imine)s: influence of their microstructures on the antimicrobial properties. Biomacromol 13(3):612–623

79. Liu SQ, Yang C, Huang Y, Ding X, Li Y, Fan WM, Hedrick JL, Yang Y-Y (2012) Antimicrobial and antifouling hydrogels formed in Situ from Polycarbonate and Poly (ethylene glycol) via Michael addition. Adv Mater 24(48):6484–6489

80. Rawlinson L-AB, SaM Ryan, Mantovani G, Syrett JA, Haddleton DM, Brayden DJ (2009) Antibacterial effects of Poly(2-(dimethylamino ethyl)methacrylate) against selected gram-positive and gram-negative bacteria. Biomacromol 11(2):443–453

81. Qiao Y, Yang C, Coady DJ, Ong ZY, Hedrick JL, Yang Y-Y (2012) Highly dynamic biodegradable micelles capable of lysing Gram-positive and Gram-negative bacterial membrane. Biomaterials 33(4):1146–1153

82. Muñoz-Bonilla A, Fernández-García M (2015) The roadmap of antimicrobial polymeric materials in macromolecular nanotechnology. Eur Polym J 65:46–62

83. Ganewatta MS, Tang C (2015) Controlling macromolecular structures towards effective antimicrobial polymers. Polymer 63:A1–A29

84. Oda Y, Kanaoka S, Sato T, Aoshima S, Kuroda K (2011) Block versus random Amphiphilic copolymers as antibacterial agents. Biomacromol 12(10):3581–3591

85. Zhu C, Yang Q, Liu L, Lv F, Li S, Yang G, Wang S (2011) Multifunctional cationic poly (p-phenylene vinylene) polyelectrolytes for selective recognition, imaging, and killing of bacteria over mammalian cells. Adv Mater 23(41):4805–4810

86. Yuan W, Wei J, Lu H, Fan L, Du J (2012) Water-dispersible and biodegradable polymer micelles with good antibacterial efficacy. Chem Commun 48(54):6857–6859

87. Tasdelen MA, Kahveci MU, Yagci Y (2011) Telechelic polymers by living and controlled/ living polymerization methods. Prog Polym Sci 36(4):455–567

88. Garcia MT, Ribosa I, Perez L, Manresa A, Comelles F (2014) Self-assembly and antimicrobial activity of long-chain amide-functionalized ionic liquids in aqueous solution. Colloids Surf B Biointerfaces 123:318–325

89. Luxenhofer R, Han Y, Schulz A, Tong J, He Z, Kabanov AV, Jordan R (2012) Poly (2-oxazoline)s as polymer therapeutics. Macromol Rapid Commun 33(19):1613–1631

90. Venkataraman S, Zhang Y, Liu L, Yang Y-Y (2010) Design, syntheses and evaluation of hemocompatible pegylated-antimicrobial polymers with well-controlled molecular structures. Biomaterials 31(7):1751–1756

91. Coady DJ, Ong ZY, Lee PS, Venkataraman S, Chin W, Engler AC, Yang YY, Hedrick JL (2014) Enhancement of cationic antimicrobial materials via cholesterol incorporation. Adv Healthcare Mater 3(6):882–889

92. Cui D, Szarpak A, Pignot-Paintrand I, Varrot A, Boudou T, Detrembleur C, Jérôme C, Picart C, Auzély-Velty R (2010) Contact-killing polyelectrolyte microcapsules based on Chitosan derivatives. Adv Funct Mater 20(19):3303–3312

93. Li LL, Xu JH, Qi GB, Zhao X, Yu F, Wang H (2014) Core-shell supramolecular gelatin nanoparticles for adaptive and on-demand antibiotic delivery. ACS Nano 8(5):4975–4983

94. Baier G, Cavallaro A, Vasilev K, Mailänder V, Musyanovych A, Landfester K (2013) Enzyme responsive hyaluronic acid nanocapsules containing polyhexanide and their exposure to bacteria to prevent infection. Biomacromol 14(4):1103–1112

95. Cao A, Tang Y, Liu Y, Yuan H, Liu L (2013) A strategy for antimicrobial regulation based on fluorescent conjugated oligomer-DNA hybrid hydrogels. Chem Commun 49(49):5574–5576

96. Noimark S, Dunnill CW, Wilson M, Parkin IP (2009) The role of surfaces in catheter-associated infections. Chem Soc Rev 38(12):3435–3448

97. Yuan Y, Sun F, Zhang F, Ren H, Guo M, Cai K, Jing X, Gao X, Zhu G (2013) Targeted synthesis of porous aromatic frameworks and their composites for versatile, facile, efficacious, and durable antibacterial polymer coatings. Adv Mater 25(45):6619–6624

98. Faure E, Falentin-Daudré C, Lanero TS, Vreuls C, Zocchi G, Van De Weerdt C, Martial J, Jérôme C, Duwez A-S, Detrembleur C (2012) Functional nanogels as platforms for imparting antibacterial, antibiofilm, and antiadhesion activities to stainless steel. Adv Funct Mater 22(24):5271–5282

99. Bai H, Yuan H, Nie C, Wang B, Lv F, Liu L, Wang S (2015) A supramolecular antibiotic switch for antibacterial regulation. Angew Chem Int Ed 54(45):13208–13213
100. Laloyaux X, Fautré E, Blin T, Purohit V, Leprince J, Jouenne T, Jonas AM, Glinel K (2010) Temperature-responsive polymer brushes switching from bactericidal to cell-repellent. Adv Mater 22(44):5024–5028
101. Yoshinari M, Oda Y, Kato T, Okuda K (2001) Influence of surface modifications to titanium on antibacterial activity in vitro. Biomaterials 22(14):2043–2048
102. Sotiriou GA, Pratsinis SE (2010) Antibacterial activity of nanosilver ions and particles. Environ Sci Technol 44(14):5649–5654
103. Lok C-N, Ho C-M, Chen R, He Q-Y, Yu W-Y, Sun H, Tam PK-H, Chiu J-F, Che C-M (2007) Silver nanoparticles: partial oxidation and antibacterial activities. J Biol Inorg Chem 12(4):527–534
104. Park MVDZ, Neigh AM, Vermeulen JP, de la Fonteyne LJJ, Verharen HW, Briedé JJ, van Loveren H, de Jong WH (2011) The effect of particle size on the cytotoxicity, inflammation, developmental toxicity and genotoxicity of silver nanoparticles. Biomaterials 32(36):9810–9817
105. Pal S, Tak YK, Song JM (2007) Does the antibacterial activity of silver nanoparticles depend on the shape of the nanoparticle? A study of the gram-negative bacterium Escherichia coli. Appl Environ Microbi 73(6):1712–1720
106. Feng QL, Wu J, Chen GQ, Cui FZ, Kim TN, Kim JO (2000) A mechanistic study of the antibacterial effect of silver ions on Escherichia coli and Staphylococcus aureus. J Biomed Mater Res 52(4):662–668
107. Maness P-C, Smolinski S, Blake DM, Huang Z, Wolfrum EJ, Jacoby WA (1999) Bactericidal activity of photocatalytic TiO$_2$ reaction: toward an understanding of its killing mechanism. Appl Environ Microbiol 65(9):4094–4098
108. Pandurangan K, Kitchen JA, Blasco S, Paradisi F, Gunnlaugsson T (2014) Supramolecular pyridyl urea gels as soft matter with antibacterial properties against MRSA and/or E. coli. Chem Commun 50 (74):10819–10822
109. Ning X, Lee S, Wang Z, Kim D, Stubblefield B, Gilbert E, Murthy N (2011) Maltodextrin-based imaging probes detect bacteria in vivo with high sensitivity and specificity. Nat Mater 10(8):602–607
110. Chung HJ, Castro CM, Im H, Lee H, Weissleder R (2013) A magneto-DNA nanoparticle system for rapid detection and phenotyping of bacteria. Nat Nanotechnol 8(5):369–375
111. Tseng Y-T, Chang H-Y, Huang C-C (2012) Mass spectrometry-based immunosensor for bacteria using antibody-conjugated gold nanoparticles. Chem Commun 48:8712–8714
112. Won BY, Yoon HC, Park HG (2008) Enzyme-catalyzed signal amplification for electrochemical DNA detection with a PNA-modified electrode. Analyst 133(1)
113. Pazos E, Sleep E, Rubert Perez CM, Lee SS, Tantakitti F, Stupp SI (2016) Nucleation and growth of ordered arrays of silver nanoparticles on peptide nanofibers: hybrid nanostructures with antimicrobial properties. J Am Chem Soc 138(17):5507–5510
114. Wang Y, Cao L, Guan S, Shi G, Luo Q, Miao L, Thistlethwaite I, Huang Z, Xu J, Liu J (2012) Silver mineralization on self-assembled peptide nanofibers for long term antimicrobial effect. J Mater Chem 22(6):2575–2581
115. Baek K, Liang J, Lim WT, Zhao H, Kim DH, Kong H (2015) In situ assembly of antifouling/ bacterial silver nanoparticle-hydrogel composites with controlled particle release and matrix softening. ACS Appl Mater Interfaces 7(28):15359–15367
116. Shome A, Dutta S, Maiti S, Das PK (2011) In situ synthesized Ag nanoparticle in self-assemblies of amino acid based amphiphilic hydrogelators: development of antibacterial soft nanocomposites. Soft Matter 7(6):3011–3022
117. Wei X, Luo M, Liu H (2014) Preparation of the antithrombotic and antimicrobial coating through layer-by-layer self-assembly of nattokinase-nanosilver complex and polyethylenimine. Colloids Surf B Biointerfaces 116:418–423

118. Agarwal A, Guthrie KM, Czuprynski CJ, Schurr MJ, McAnulty JF, Murphy CJ, Abbott NL (2011) Polymeric multilayers that contain silver nanoparticles can be stamped onto biological tissues to provide antibacterial activity. Adv Funct Mater 21(10):1863–1873

119. Jia Q, Shan S, Jiang L, Wang Y, Li D (2012) Synergistic antimicrobial effects of polyaniline combined with silver nanoparticles. J Appl Polym Sci 125(5):3560–3566

120. Lu H, Fan L, Liu QM, Wei JR, Ren TB, Du JZ (2012) Preparation of water-dispersible silver-decorated polymer vesicles and micelles with excellent antibacterial efficacy. Polym Chem 3(8):2217–2227

121. Ocsoy I, Paret ML, Ocsoy MA, Kunwar S, Chen T, You M, Tan W (2013) Nanotechnology in plant disease management: DNA-directed silver nanoparticles on graphene oxide as an antibacterial against Xanthomonas perforans. ACS Nano 7(10):8972–8980

122. Brahmachari S, Mandal SK, Das PK (2014) Fabrication of SWCNT-Ag nanoparticle hybrid included self-assemblies for antibacterial applications. PLoS One 9(9):e106775

123. Lv M, Su S, He Y, Huang Q, Hu W, Li D, Fan C, Lee S-T (2010) Long-term antimicrobial effect of silicon nanowires decorated with silver nanoparticles. Adv Mater 22(48):5463–5467

124. Ruparelia JP, Chatterjee AK, Duttagupta SP, Mukherji S (2008) Strain specificity in antimicrobial activity of silver and copper nanoparticles. Acta Biomater 4(3):707–716

125. Mahmoudi M, Serpooshan V (2012) Silver-coated engineered magnetic nanoparticles are promising for the success in the fight against antibacterial resistance threat. ACS Nano 6(3):2656–2664

126. Wang L, Luo J, Shan S, Crew E, Yin J, Zhong C-J, Wallek B, Wong SSS (2011) Bacterial inactivation using silver-coated magnetic nanoparticles as functional antimicrobial agents. Anal Chem 83(22):8688–8695

127. Pang M, Hu J, Zeng HC (2010) Synthesis, morphological control, and antibacterial properties of hollow/solid Ag2S/Ag heterodimers. J Am Chem Soc 132(31):10771–10785

128. Alonso A, Muñoz-Berbel X, Vigués N, Rodríguez-Rodríguez R, Macanás J, Muñoz M, Mas J, Muraviev DN (2013) Superparamagnetic Ag@Co-Nanocomposites on granulated cation exchange polymeric matrices with enhanced antibacterial activity for the environmentally safe purification of water. Adv Funct Mater 23(19):2450–2458

129. Xu J, Zhou X, Gao Z, Song Y-Y, Schmuki P (2016) Visible-light-triggered drug release from $TiO_2$ nanotube arrays: a controllable antibacterial platform. Angew Chem Int Ed 55(2):593–597

130. Gehring J, Trepka B, Klinkenberg N, Bronner H, Schleheck D, Polarz S (2016) Sunlight-triggered nanoparticle synergy: teamwork of reactive oxygen species and nitric oxide released from mesoporous organosilica with advanced antibacterial activity. J Am Chem Soc 138(9):3076–3084

131. Qiu Q, Liu T, Li Z, Ding X (2015) Facile synthesis of N-halamine-labeled silica-polyacrylamide multilayer core-shell nanoparticles for antibacterial ability. J Mater Chem B 3(36):7203–7212

132. Fei J, Zhao J, Du C, Wang A, Zhang H, Dai L, Li J (2014) One-pot ultrafast self-assembly of autofluorescent polyphenol-based core@shell nanostructures and their selective antibacterial applications. ACS Nano 8(8):8529–8536

133. Pal S, Yoon EJ, Tak YK, Choi EC, Song JM (2009) Synthesis of highly antibacterial nanocrystalline trivalent silver polydiguanide. J Am Chem Soc 131(44):16147–16155

134. Zhou B, Li Y, Deng H, Hu Y, Li B (2014) Antibacterial multilayer films fabricated by layer-by-layer immobilizing lysozyme and gold nanoparticles on nanofibers. Colloids Surf B Biointerfaces 116:432–438

135. L-l Li, Wang H (2013) Enzyme-coated mesoporous silica nanoparticles as efficient antibacterial agents in vivo. Adv Healthcare Mater 2(10):1351–1360

136. Liu L, Yang J, Xie J, Luo Z, Jiang J, Yang YY, Liu S (2013) The potent antimicrobial properties of cell penetrating peptide-conjugated silver nanoparticles with excellent selectivity for gram-positive bacteria over erythrocytes. Nanoscale 5(9):3834–3840

137. Taglietti A, Diaz Fernandez YA, Amato E, Cucca L, Dacarro G, Grisoli P, Necchi V, Pallavicini P, Pasotti L, Patrini M (2012) Antibacterial activity of glutathione-coated silver nanoparticles against gram positive and gram negative bacteria. Langmuir 28(21):8140–8148

138. Hsiao C-W, Chen H-L, Liao Z-X, Sureshbabu R, Hsiao H-C, Lin S-J, Chang Y, Sung H-W (2015) Effective photothermal killing of pathogenic bacteria by using spatially tunable colloidal gels with nano-localized heating sources. Adv Funct Mater 25(5):721–728

139. Borovička J, Metheringham WJ, Madden LA, Walton CD, Stoyanov SD, Paunov VN (2013) Photothermal colloid antibodies for shape-selective recognition and killing of microorganisms. J Am Chem Soc 135(14):5282–5285

140. Majumdar P, Nomula R, Zhao J (2014) Activatable triplet photosensitizers: magic bullets for targeted photodynamic therapy. J Mater Chem C 2(30):5982–5997

141. Barroso Á, Grüner M, Forbes T, Denz C, Strassert CA (2016) Spatiotemporally resolved tracking of bacterial responses to ROS-mediated damage at the single-cell level with quantitative functional microscopy. ACS Appl Mater Interfaces 8(24):15046–15057

142. Li LL, Ma HL, Qi GB, Zhang D, Yu F, Hu Z, Wang H (2016) Pathological-condition-driven construction of supramolecular nanoassemblies for bacterial infection detection. Adv Mater 28(2):254–262

143. Thomas M, Craik JD, Tovmasyan A, Batinic-Haberle I, Benov LT (2015) Amphiphilic cationic Zn-porphyrins with high photodynamic antimicrobial activity. Future Microbiol 10 (5):709–724

144. Johnson GA, Muthukrishnan N, Pellois J-P (2012) Photoinactivation of gram positive and gram negative bacteria with the antimicrobial peptide (KLAKLAK)2 Conjugated to the hydrophilic photosensitizer Eosin Y. Bioconjug Chem 24(1):114–123

145. Zhou X, Chen Z, Wang Y, Guo Y, Tung C-H, Zhang F, Liu X (2013) Honeycomb-patterned phthalocyanine films with photo-active antibacterial activities. Chem Commun 49 (90):10614–10616

146. Liu K, Liu Y, Yao Y, Yuan H, Wang S, Wang Z, Zhang X (2013) Supramolecular photosensitizers with enhanced antibacterial efficiency. Angew Chem Int Ed 52(32):8285–8289

147. Shrestha A, Kishen A (2012) Polycationic Chitosan-conjugated photosensitizer for antibacterial photodynamic therapy†. Photochem Photobiol 88(3):577–583

148. Chong H, Nie C, Zhu C, Yang Q, Liu L, Lv F, Wang S (2011) Conjugated polymer nanoparticles for light-activated anticancer and antibacterial activity with imaging capability. Langmuir 28(4):2091–2098

149. Yang K, Gitter B, Rüger R, Albrecht V, Wieland GD, Fahr A (2012) Wheat germ agglutinin modified liposomes for the photodynamic inactivation of bacteria†. Photochem Photobiol 88 (3):548–556

150. Shijie L, Shenglin Q, Lili L, Guobin Q, Yaoxin L, Zengying Q, Hao W, Chen S (2015) Surface charge-conversion polymeric nanoparticles for photodynamic treatment of urinary tract bacterial infections. Nanotechnology 26(49):495602

151. Xiong M-H, Li Y-J, Bao Y, Yang X-Z, Hu B, Wang J (2012) Bacteria-responsive multifunctional nanogel for targeted antibiotic delivery. Adv Mater 24(46):6175–6180

152. Xiong M-H, Bao Y, Yang X-Z, Wang Y-C, Sun B, Wang J (2012) Lipase-sensitive polymeric triple-layered nanogel for "on-demand" drug delivery. J Am Chem Soc 134 (9):4355–4362

153. Shukla A, Fang JC, Puranam S, Hammond PT (2012) Release of vancomycin from multilayer coated absorbent gelatin sponges. J Control Release 157(1):64–71

154. Choi SK, Myc A, Silpe JE, Sumit M, Wong PT, McCarthy K, Desai AM, Thomas TP, Kotlyar A, Holl MMB, Orr BG, Baker JR (2012) Dendrimer-based multivalent vancomycin nanoplatform for targeting the drug-resistant bacterial surface. ACS Nano 7(1):214–228

155. Ren C, Wang H, Zhang X, Ding D, Wang L, Yang Z (2014) Interfacial self-assembly leads to formation of fluorescent nanoparticles for simultaneous bacterial detection and inhibition. Chem Commun 50(26):3473–3475

156. Sémiramoth N, Meo CD, Zouhiri F, Saïd-Hassane F, Valetti S, Gorges R, Nicolas V, Poupaert JH, Chollet-Martin S, Desmaële D, Gref R, Couvreur P (2012) Self-assembled squalenoylated penicillin bioconjugates: an original approach for the treatment of intracellular infections. ACS Nano 6(5):3820–3831

157. Guchhait G, Altieri A, Gorityala B, Yang X, Findlay B, Zhanel GG, Mookherjee N, Schweizer F (2015) Amphiphilic tobramycins with immunomodulatory properties. Angew Chem Int Ed 54(21):6278–6282

158. Ray PC, Khan SA, Singh AK, Senapati D, Fan Z (2012) Nanomaterials for targeted detection and photothermal killing of bacteria. Chem Soc Rev 41(8):3193–3209

159. Wu J, Zawistowski A, Ehrmann M, Yi T, Schmuck C (2011) Peptide functionalized polydiacetylene liposomes act as a fluorescent turn-on sensor for bacterial lipopolysaccharide. J Am Chem Soc 133(25):9720–9723

160. Miranda OR, Li X, Garcia-Gonzalez L, Zhu Z-J, Yan B, Bunz UHF, Rotello VM (2011) Colorimetric bacteria sensing using a supramolecular enzyme-nanoparticle biosensor. J Am Chem Soc 133(25):9650–9653

161. Saneyoshi H, Ito Y, Abe H (2013) Long-lived luminogenic probe for detection of RNA in a crude solution of living bacterial cells. J Am Chem Soc 135(37):13632–13635

162. Gao J, Li L, Ho PL, Mak GC, Gu H, Xu B (2006) Combining fluorescent probes and biofunctional magnetic nanoparticles for rapid detection of bacteria in human blood. Adv Mater 18(23):3145–3148

163. Qi G, Li L, Yu F, Wang H (2013) Vancomycin-modified mesoporous silica nanoparticles for selective recognition and killing of pathogenic gram-positive bacteria over macrophage-like cells. ACS Appl Mater Interfaces 5(21):10874–10881

164. Azzopardi EA, Ferguson EL, Thomas DW (2013) The enhanced permeability retention effect: a new paradigm for drug targeting in infection. J Antimicrob Chemother 68(2):257–274

165. Agard NJ, Maltby D, Wells JA (2010) Inflammatory stimuli regulate caspase substrate profiles. Mol Cell Proteom 9(5):880–893

166. Isakov O, Modai S, Shomron N (2011) Pathogen detection using short-RNA deep sequencing subtraction and assembly. Bioinformatics 27(15):2027–2030

167. Milo S, Thet NT, Liu D, Nzakizwanayo J, Jones BV, Jenkins ATA (2016) An in-situ infection detection sensor coating for urinary catheters. Biosens Bioelectron 81:166–172

168. Izadi Z, Sheikh-Zeinoddin M, Ensafi AA, Soleimanian-Zad S (2016) Fabrication of an electrochemical DNA-based biosensor for Bacillus cereus detection in milk and infant formula. Biosens Bioelectron 80:582–589

169. Braiek M, Rokbani KB, Chrouda A, Mrabet B, Bakhrouf A, Maaref A, Jaffrezic-Renault N (2012) An electrochemical immunosensor for detection of Staphylococcus aureus bacteria based on immobilization of antibodies on self-assembled monolayers-functionalized gold electrode. Biosensors 2(4):417–426

170. Dudak FC, Boyaci İH (2014) Peptide-based surface plasmon resonance biosensor for detection of staphylococcal enterotoxin B. Food Anal Methods 7(2):506–511

171. Chen K, Wu L, Jiang X, Lu Z, Han H (2014) Target triggered self-assembly of Au nanoparticles for amplified detection of Bacillus thuringiensis transgenic sequence using SERS. Biosens Bioelectron 62:196–200

172. Xiang Y, Zhang H, Jiang B, Chai Y, Yuan R (2011) Quantum dot layer-by-layer assemblies as signal amplification labels for ultrasensitive electronic detection of uropathogens. Anal Chem 83(11):4302–4306

173. Budin G, Chung HJ, Lee H, Weissleder R (2012) A magnetic gram stain for bacterial detection. Angew Chem Int Ed 51(31):7752–7755

174. Chung HJ, Reiner T, Budin G, Min C, Liong M, Issadore D, Lee H, Weissleder R (2011) Ubiquitous detection of gram-positive bacteria with bioorthogonal magnetofluorescent nanoparticles. ACS Nano 5(11):8834–8841

175. Kinnunen P, Carey ME, Craig E, Brahmasandra SN, McNaughton BH (2014) Rapid bacterial growth and antimicrobial response using self-assembled magnetic bead sensors. Sens Actuat B Chem 190:265–269
176. Xu Y-G, Liu Z-M, Zhang B-Q, Qu M, Mo C-S, Luo J, Li S-L (2016) Development of a novel target-enriched multiplex PCR (Tem-PCR) assay for simultaneous detection of five foodborne pathogens. Food Control 64:54–59
177. Yang K, Jenkins DM, Su WW (2011) Rapid concentration of bacteria using submicron magnetic anion exchangers for improving PCR-based multiplex pathogen detection. J Microbiol Methods 86(1):69–77
178. Leevy WM, Gammon ST, Jiang H, Johnson JR, Maxwell DJ, Jackson EN, Marquez M, Piwnica-Worms D, Smith BD (2006) Optical imaging of bacterial infection in living mice using a fluorescent near-infrared molecular probe. J Am Chem Soc 128(51):16476–16477
179. van Oosten M, Schafer T, Gazendam JA, Ohlsen K, Tsompanidou E, de Goffau MC, Harmsen HJ, Crane LM, Lim E, Francis KP, Cheung L, Olive M, Ntziachristos V, van Dijl JM, van Dam GM (2013) Real-time in vivo imaging of invasive- and biomaterial-associated bacterial infections using fluorescently labelled vancomycin. Nat Commun 4:2584
180. Dumarey N, Egrise D, Blocklet D, Stallenberg B, Remmelink M, del Marmol V, Van Simaeys G, Jacobs F, Goldman S (2006) Imaging infection with 18F-FDG–labeled leukocyte PET/CT: initial experience in 21 patients. J Nucl Med 47(4):625–632
181. Clinical diagnosis of bacterial infection via FDG-PET imaging (2013) Can Chem Trans 1 (2):85–104
182. Singh A, Arutyunov D, Szymanski CM, Evoy S (2012) Bacteriophage based probes for pathogen detection. Analyst 137(15):3405–3421
183. Edgar R, McKinstry M, Hwang J, Oppenheim AB, Fekete RA, Giulian G, Merril C, Nagashima K, Adhya S (2006) High-sensitivity bacterial detection using biotin-tagged phage and quantum-dot nanocomplexes. Proc Natl Acad Sci USA 103:4841–4845
184. Zhao X, Hilliard LR, Mechery SJ, Wang Y, Bagwe RP, Jin S, Tan W (2004) A rapid bioassay for single bacterial cell quantitation using bioconjugated nanoparticles. Proc Natl Acad Sci USA 101(42):15027–15032
185. Thet NT, Alves DR, Bean JE, Booth S, Nzakizwanayo J, Young AER, Jones BV, Jenkins ATA (2016) Prototype development of the intelligent hydrogel wound dressing and its efficacy in the detection of model pathogenic wound biofilms. ACS Appl Mater Interfaces 8(24):14909–14919
186. Zhou J, Loftus AL, Mulley G, Jenkins ATA (2010) A thin film detection/response system for pathogenic bacteria. J Am Chem Soc 132(18):6566–6570
187. da Silva JSL, Oliveira MDL, de Melo CP, Andrade CAS (2014) Impedimetric sensor of bacterial toxins based on mixed (Concanavalin A)/polyaniline films. Colloids Surf B Biointerfaces 117:549–554
188. Haas S, Hain N, Raoufi M, Handschuh-Wang S, Wang T, Jiang X, Schönherr H (2015) Enzyme degradable polymersomes from hyaluronic acid-block-poly($\varepsilon$-caprolactone) copolymers for the detection of enzymes of pathogenic bacteria. Biomacromol 16(3):832–841
189. Tücking K-S, Grützner V, Unger RE, Schönherr H (2015) Dual enzyme-responsive capsules of hyaluronic acid-block-Poly(Lactic Acid) for sensing bacterial enzymes. Macromol Rapid Commun 36(13):1248–1254
190. Mouffouk F, da Costa AMR, Martins J, Zourob M, Abu-Salah KM, Alrokayan SA (2011) Development of a highly sensitive bacteria detection assay using fluorescent pH-responsive polymeric micelles. Biosens Bioelectron 26(8):3517–3523
191. Guo Y, Wang Y, Liu S, Yu J, Wang H, Cui M, Huang J (2015) Electrochemical immunosensor assay (EIA) for sensitive detection of E. coli O157:H7 with signal amplification on a SG-PEDOT-AuNPs electrode interface. Analyst 140(2):551–559
192. Maurer EI, Comfort KK, Hussain SM, Schlager JJ, Mukhopadhyay SM (2012) Novel platform development using an assembly of carbon nanotube, nanogold and immobilized RNA capture element towards rapid selective sensing of bacteria. Sensors 12(6):8135

193. Lian Y, He F, Wang H, Tong F (2015) A new aptamer/graphene interdigitated gold electrode piezoelectric sensor for rapid and specific detection of staphylococcus aureus. Biosens Bioelectron 65:314–319
194. Wan Y, Lin Z, Zhang D, Wang Y, Hou B (2011) Impedimetric immunosensor doped with reduced graphene sheets fabricated by controllable electrodeposition for the non-labelled detection of bacteria. Biosens Bioelectron 26(5):1959–1964
195. Chang J, Mao S, Zhang Y, Cui S, Zhou G, Wu X, Yang C-H, Chen J (2013) Ultrasonic-assisted self-assembly of monolayer graphene oxide for rapid detection of Escherichia coli bacteria. Nanoscale 5(9):3620–3626
196. Ding X, Li H, Deng L, Peng Z, Chen H, Wang D (2011) A novel homogenous detection method based on the self-assembled DNAzyme labeled DNA probes with SWNT conjugates and its application in detecting pathogen. Biosens Bioelectron 26(11):4596–4600
197. Kim I, Jeong H-H, Kim Y-J, Lee N-E, K-m Huh, Lee C-S, Kim GH, Lee E (2014) A Light-up 1D supramolecular nanoprobe for silver ions based on assembly of pyrene-labeled peptide amphiphiles: cell-imaging and antimicrobial activity. J Mater Chem B 2(38):6478–6486
198. Gao M, Hu Q, Feng G, Tomczak N, Liu R, Xing B, Tang BZ, Liu B (2015) A Multifunctional probe with aggregation-induced emission characteristics for selective fluorescence imaging and photodynamic killing of bacteria over mammalian cells. Adv Healthcare Mater 4(5):659–663
199. Franzini RM, Kool ET (2011) Improved templated fluorogenic probes enhance the analysis of closely related pathogenic bacteria by microscopy and flow cytometry. Bioconjug Chem 22(9):1869–1877
200. Deng B, Chen J, Zhang H (2014) Assembly of multiple DNA components through target binding toward homogeneous, isothermally amplified, and specific detection of proteins. Anal Chem 86(14):7009–7016
201. Eker B, Yilmaz MD, Schlautmann S, Gardeniers JGE, Huskens J (2011) A supramolecular sensing platform for phosphate anions and an anthrax biomarker in a microfluidic device. Int J Mol Sci 12(11):7335
202. Ning X, Lee S, Wang Z, Kim D, Stubblefield B, Gilbert E, Murthy N (2011) Maltodextrin-based imaging probes detect bacteria in vivo with high sensitivity and specificity. Nat Mater 10(8):602–607
203. Panizzi P, Nahrendorf M, Figueiredo J-L, Panizzi J, Marinelli B, Iwamoto Y, Keliher E, Maddur AA, Waterman P, Kroh HK, Leuschner F, Aikawa E, Swirski FK, Pittet MJ, Hackeng TM, Fuentes-Prior P, Schneewind O, Bock PE, Weissleder R (2011) In vivo detection of Staphylococcus aureus endocarditis by targeting pathogen-specific prothrombin activation. Nat Med 17(9):1142–1146
204. Zhang D, Qi GB, Zhao YX, Qiao SL, Yang C, Wang H (2015) In situ formation of nanofibers from purpurin18-peptide conjugates and the assembly induced retention effect in tumor sites. Adv Mater 27(40):6125–6130

# Chapter 4
# Enzyme-Instructed Self-assembly of Small Peptides In Vivo for Biomedical Application

**Zhentao Huang and Yuan Gao**

**Abstract** With the development of technology, there have developed many methods to treat diseases. Among them, precision medicine is in an urgent need for public healthcare. In the past several decades, the rapid development in nanotechnology significantly improves the realization of precision nanomedicine. Comparing to well-established nanoparticle-based strategy, in this chapter, we focus on the strategy using enzyme-instructed self-assembly (EISA) in biological milieu for biomedical application. Generally speaking, the principles of designing small molecules for EISA require two aspects: (1) the substrate of enzyme of interest; (2) self-assembly potency after enzymatic conversion. This strategy has shown its irreplaceable advantages in nanomedicine, specific for cancer treatments and Vaccine Adjuvants. Up to now, all the reported examples rely on only one kind of enzyme-hydrolase. Therefore, we envision that the application of EISA strategy just begins and will lead a new paradigm in nanomedicine.

**Keywords** Nanotechnology · Enzyme · Self-assembly · Biomedical application

## 4.1 Introduction

Supramolecular chemistry focuses on the intermolecular bond and the structures and functions of the supramolecules, while the molecular chemistry is based on the covalent bond [1]. The molecular self-assembly is a branch of supramolecular chemistry [2]. The macrobehavior of molecular self-assembly is to form gels.

Z. Huang · Y. Gao (✉)
Beijing Engineering Research Center for Bionanotechnology and CAS Key Laboratory for Biomedical Effects of Nanomaterials and Nanosafety, CAS Center for Excellence in Nanoscience, National Center for Nanoscience and Technology, No. 11 Zhongguancun Beiyitiao, Beijing 100190, People's Republic of China
e-mail: gaoy@nanoctr.cn

Z. Huang
e-mail: huangzt@nanoctr.cn

© Springer Nature Singapore Pte Ltd. 2018
H. Wang and L.-L. Li (eds.), *In Vivo Self-Assembly Nanotechnology for Biomedical Applications*, Nanomedicine and Nanotoxicology,
https://doi.org/10.1007/978-981-10-6913-0_4

Although molecular gels have been studied for over 170 years, the discovery and design of small molecules forming gel have still drawn great attention of many scientists due to their potential applications in tissue engineering drug release [3, 4] and drug release [5, 6]. As we all know, for biological applications, one of the most important problems is to find biodegradable materials to realize controlled drug release. Compared to polymers gels, the small molecule gels could resolve this problem because that they consist of biocompatible components and are held together by noncovalent force, which make them easier for the body to degrade. Therefore, the research on the small molecule gels is rapidly developing.

## 4.2   The Development of Small Molecular Gels

According to Flory, a gel has a continuous structure with macroscopic dimensions that is permanent on the time scale of an analytical experiment and is solid-like in its rheological behavior below a certain stress limit. In another words, the gels consist of two parts, the gelators and the solvent. Generally speaking, the gels have quite a few gelators, usually within the range of 0.1–10 wt%. When the gel formed, in the macroscope, the solvent is immobilized and can support its own weight and is not flowed. The simplest test is to turn the test tube upside down and observe it flows or not. In the microscope, the gel forms many nanofibers, which are twisted and traps the solvent by surface tension [7, 8]. To molecular gels, there already have many studies on mechanisms of gelators, [9, 10] different types of gelator, [11, 12], and their applications [13, 14]. Considering the difference of the solvents, the gels can be simply classified into organogels and hydrogels, which have different applications.

### 4.2.1   The Development and Characteristic of Organogels

The organogels are called "π-gelators" which are soft materials. The "π-gelators" derive from gelators with more than one aromatic π-unit, which has contributed to self-assembly and easier to from gels. Besides, duo to their delocalized π-electrons, π-systems often have some special properties such as electronic conductivity and luminescence [15, 16]. The π-systems have great applications in organic electronic devices by coating if the size and shape of aggregates can be controlled. And one possible way is to use the weak force such as hydrogen bonding and π-stacking interactions to realize self-assembly. In organogels, the solvents can almost be any common organic solvents as far as recent studies. Simply speaking, gelation is a balance between crystallization and solubilization. Therefore, to a given solvent, a molecule requires functionality that will provide both of them. Mostly, the gelators of organogels have some functional groups such as hydroxyl and carboxylic acid, which has been proved to be essential for gelation because that they are capable of forming hydrogen bond. And an aromatic core is another characteristic of organ-gelators because the presence of an aromatic core can help $\pi-\pi$ stacking. The organogels can be triggered by light, pH, temperature, ionic strength, and other factors.

Just like many scientific discoveries, we are not planned to enter into the field of organogels. It was found by Y.-C. Lin during a photochemical investigation [17]. In this experiment, the 3-b-cholesteryl-4-(2-anthryloxy) butanoate (CAB) can turn organic solution to gel at very low concentrations (less than 2 wt%). Then, the scientists focus on searching for the simpler organogelators. The molecules with only one aromatic group and one linking chain had achieved little success. And the molecules with one aromatic group and two linking chain has been proved to successfully form organogels in many organic solutions [18]. And with the developing of the investigation, some conclusion has been drawn:

(1) H-bond is not inevitable if other packing contribution dominated such as $\pi-\pi$ interactions and London dispersion forces [19].
(2) the interactions of charge—transfer among gelators can help to stabilize gels [20].
(3) thixotropy can be induced by adding a small concentration of a second mole-cule with poor ability to form gel, which is similar to the gelator in structure.
(4) the temperature and solubility of the gelator in organic solvent have great influence on the fraction of the gelator within the gels [21].
(5) what determines the shape, the dimension, and the $T_{gel}$ (the sol↔gel transition temperature) values are the properties of the liquid mixture, rather than the individual components [22].

## 4.2.2 The Development and Characteristic of Hydrogels

With the development of organogels, the limitation is obvious when applied to biomedical field. And water is the unique solvent to maintain life forms and is the most abundant substance in the body of life. In 1921, Hoffman reported a first small molecule hydrogelator, the compound dibenzoyl-L-cystine, which could form gel at 0.1 wt% concentration in macroscope [23]. And about 100 years later, Menger and co-workers applied modern physical methods such as X-ray crystallography and electron microscope to examine the hydrogel of dibenzoyl-L-cystine [24]. And they revealed the microstructure of hydrogels. Besides, they found that aromatic moieties have great improvement on intermolecular interaction in water. Although there are many principles of supramolecular hydrogelation, the first gel is an accidental dis-covery of a particular molecule that forms a gel in a solvent. Just like organogels, it usually gains hydrogels of peptides in accident when the researchers invest the oli-gomeric peptides [6, 25]. And it can be concluded that the small molecules self-assembly is a universal phenomenon, rather than particular process. Therefore, it will be meaningful to explore the supramolecular hydrogels. Essentially, the hydrogels formed by the self-assembly of small organic molecules are different from organogels and polymer hydrogels. The polymeric hydrogels originated from a random cross-linked network is a strong covalent bond. As to hydrogels, the molecular self-assembly is forced by weak and noncovalent interaction among hydrogelators in water, which make the hydrogels more ordered. While simple swelling usually makes a polymeric hydrogel, a stimulus or a triggering force is necessary to bias thermodynamic equilibrium for initiating the self-assembly process or phase transition to obtain a supramolecular hydrogel. Therefore, there are many forms of stimuli or triggers for manipulating the weak interactions. The methods can be divided into physical methods (such as changing the temperature, applying ultrasound, or modulating the ionic strength) and the chemical methods (such as pH change, chemical or photochemical reactions, redox, and catalysis). Among them, the change of pH is the simplest method to gain hydrogel because a small amount of acid or base easily and rapidly can lead to a large pH shift via a diffusion-limited process. However, considering the biomedicine applications, the methods of great change of the environment (such as all the physical methods and the change of pH) may be useless. And the methods based on chemical reaction are promising. Chemical reactions can yield new products which have great different properties from reactants and turn the solution to gel. In polymeric hydrogels, click chemistry, [26] redox reactions, [27, 28] Michael reaction, [29] acid–base reaction [30], and ligation reaction [31] have already been developed. To supramolecular hydrogels, there are still much less attention. Considering that self-assembly is the molecular foundation of life, and soft and wet are another two obvious characteristics of most types of cells, it is not surprising that catalysis and enzymes are attracting increased attention and are achieving many unexpected successes in the generation and applications of supramolecular hydrogels. The typical experiment is reported by Xu group [30]. They synthesized the hydrogelator with good water solubility by the replacement of a carboxylate group with a hydroxyl group. By the hydrolysis of the carboxylic ester bond, it turned to be high hydrophobic and formed the gels, which is much stable over a wide pH(1–14) range. And this may be a possible application in designing a robust system of prodrug. Then, the group of Hamachi reported that the "retro-Diels-Alder" reaction can be used to trigger morphological transformation of supramolecular nanostructures, which triggered by heating [32]. Due to the complex environment in cells, the enzyme has greater advantages than other triggers on self-assembly.

### 4.2.3   The Development of Enzyme-Instructed Self-assembly

In 2015, the announcement of Precision Medicine Initiative (PMI) envisioned the new era of medicine to develop individualized care [33]. The establishment of PMI indicates that there are still many incurable or lack of effective treatment for a lot of diseases, especially cancer. The improvement in cancer treatment is calling for the urgent advances in new technology, such as nanotechnology. About three decades ago, the blossoming bionanotechnology has made remarkable progresses in precision medicine [34–36]. When it starts, in this field, the major efforts are to fabricate multi-functional nanoparticles to realize diagnostic and therapeutic effect [37]. The vital concept is to carry out and maximize the targeting capability of nanoparticles while leaving the rest of body unaffected [38, 39]. Up to now, nanoparticle based on cancer theranostics has made tremendous progress which has been thoroughly reviewed [40–42]. Nevertheless, the nanoparticle-based strategy still remains in a dilemma. Before real application in market, there are a list of unavoidable issues to be solved, mainly including (1) the quality control of multi-step fabrication process; (2) the difficulty in clearance of nanoparticle; (3) the poor ability of penetration into the tumor because of the size of nanoparticle; (4) the high risky off target on account of frequent mutation of cancer cell surface proteins, and (5) the delivery efficiency of nanoparticle [43–45]. In contrast, the bottom-up strategy, which can generate nanomaterial directly in cancer cells rather than delivery the nanomaterial to cancer cells, has drawn more and more attentions in recent ten years [46]. Apart from the achievement of the in vivo formation of quantum dots for bio-applications, [47] one kind of nanomaterial which uses enzyme to activate self-assembly process of small molecules [48] has already been demonstrated the great potential applications ranging from cancer diagnosis to therapy or their combination (so-called theranostics). In this chapter, we summarize the recent progress on the EISA and classify the enzymes underlying enzymatic reactions in details. At the same time, if the readers have interests in various application of EISA, there are a few excellent reviews on the topic of supramolecular self-assembly for potential anticancer therapeutics for further reading, [49–51] which emphasize on supramolecular hydrogels as molecular biomaterial, the intersection of supramolecular chemistry, biomedicine science, and the biological functions of prion-like nanofibrils of small molecules, respectively. We just pay our attention to the biomedical application of EISA based on the difference of enzymes.

## 4.3 The Characteristic and Advantages of EISA on Cancer Theranostics

EISA is the multi-step process to form special functional structures, in which amphiphilic molecules undergo certain transformations usually triggered by enzymes before they stack with each other [48]. As a ruler, the formation of self-assembled objects can be kinetically controlled, which may be potential application on control drug release. In addition, there is still emerging example under mechanical control directed by catalytic action [52] or non-equilibrium transient state [53]. These structures of hydrogels display a variety of micromorphologies such as nanoparticles, nanofibers, ribbons, and so on [54–56]. And the different morphologies depend on the specific geometry of certain building blocks–molecular structures. Although it is different morphologies, all of them share some same features, such as the intermolecular interactions which are noncovalent including hydrogen bonding, aromatic–aromatic interaction, electrostatic attraction, and other weak Van der Waals forces [57]. For example, the hydrogen bonding and aromatic-aromatic interaction are directional force, which eventually determine the diameter of nanoparticles, the infinite length along the prolate axis, well-defined width of nanofibers or tubes, repeated unit with fixed length in helical ribbons [58]. Among the various process of self-assembly, the unique feature of EISA is the involvement of enzymatic reactions [59]. In other words, all of the EISA is triggered by enzymes, and the difference lies in the types of enzyme. Enzyme-instructed self-assembly is quite common phenomenon but much critical process in living systems. As we known, the enzyme-catalyzed processes can regulate the formation of intracellular vesicles, the dynamics of cellular skeletons formed by F-actins and microtubules (MT). Further, the dynamic cellular transformations are largely dependent on enzymatic regulations, such as the disassembly and reformation of cell membrane, F-actin and MT during the mitosis, cell movement, and so on [60, 61]. Inspired by nature, the strategy using enzyme to instruct self-assembly to synthesize small molecules has already been developed a decade ago [62]. According to the reported documents, it is necessary to design a suitable hydrophilic substrate which is so-called precursor, just like as hydrogelator. By means of enzymatic reaction, such a precursor can be easily turned into corresponding amphiphilic form [63]. The generating amphiphilic molecules will self-assemble once its concentration reached the critical assembly/aggregation concentration (CAC) and yield various nanostructures eventually.

Considering the application in living system, EISA is irreplaceable process, specifically for mammals. For mammals, the basic physiological conditions remain almost constant. Therefore, this situation is improper to adopt usual sol-gel transition, which generally requires external physical or chemical stimuli including pH, temperature, and ionic strength. Comparing to them, EISA is not only the isothermal process but also performs in physiological condition, which means not necessary to change the environment. Based on these biocompatible features, the establishment of EISA makes it has great advantages to construct self-assemblies in biological milieu.

The main advantages of the application of EISA in cancer theranostics can be summarized as follows:

(1) The material can be easily synthesized by regular chemical methods. The precursor is mainly peptides and the derivative of peptides. And the different functional amino acid can be easily purchased by commercial channels. Then, the synthesis of peptides depends on the solid phase peptide synthesis (SPPS), which can be easily obtained at a large scale with high purity. Besides, the peptides can be simply modified because of the existence of the amino and carboxyl groups.

(2) The clearance can simply realize. Because the formation of the material is based on the noncovalent interactions, the self-assembly process can be reversible to disassemble to release the original small molecules into biological environment, which get cleared via reticuloendothelial system (RES).

(3) The adequate penetration into the tumor can be got because of the size of precursor. Compared to the shallow penetration of nanoparticle into tumor, which is limited by interstitial pressure, it is much easier for small molecules (precursor) to penetrate deep inside of tumor and even tumor cells by means of passive diffusion or active transportation. Besides, there are more choices to design the appropriate precursor to get better ability of penetration into tumor.

(4) The delivery efficiency is quite sufficient. The EISA depends on the over-expression of certain enzyme rather than cell surface receptor or vascular leakage, which indicate that it can increase the opportunity of entering cells. Because that overexpressed enzyme is the intrinsic component in tumor cells, ideally, the enzyme-instructed self-assembly will proceed only in tumor cells and get small molecules accumulated. When the concentration is over CAC, the intracellular concentration of small molecules would reach millimolar level and the self assembly occurs, which can satisfy the need of preloaded or post-delivered therapeutic agents.

(5) Further, this strategy owns the potential to target "undruggable" targets or "untargetable" features of cancer cells and provides opportunities for simultaneously interacting with multiple targets [48]. EISA can greatly enrich the ways of release of drugs and enhance the solubility of some drugs.

Although there are as many as thousand kinds of enzymes in cells, the types of enzyme to trigger self-assembly is limited. Here in this chapter, we would like to elucidate the progress on the enzyme-instructed self assembly in vivo for biomedicine application in the order of enzyme.

## 4.4 The Application of EISA Strategy in Cancer Theranostics

The EISA of small molecules has been proven to be a promising method in selective inhibition of cancer cells. However, just as everything has two sides, the EISA also have some problems and limitations to solve. One of the problems is that the inhibitory concentrations of those self-assembling molecules remain too high compared to traditional pharmaceuticals due to its high CAC [64]. And lack of understanding of the interaction between potential protein targets and the in situ assemblies or aggregates, as well as the limited techniques to identify and characterize the interactions between nanoscale assemblies of small molecules and proteins, remained another inevitable obstacle for further advances of EISA [65]. Therefore, it demands more researchers and scientists to work on it. The application of this concept in biomedicine was just getting started. Next, we introduce in detail the supramolecular self-assembly induced by different kinds of enzymes. For the convenience of readers, we summarize some important parameters of EISA in living cells and animals, such as concentration and incubation time in recent reported documents to more intuitively compare with each other. (Table 4.1)

**Table 4.1** EISA in living cells and animals

| Compound | | 1 | 2 | 3ª | 4ᵇ | 5ᶜ | 6 | 7 | 7 | 8 | 9 | 10 | 11ᶠ |
|---|---|---|---|---|---|---|---|---|---|---|---|---|---|
| Cell | Conc./μM | 500 | 500 | 500 | 500 | 500 | 500 | 500 | 500 | 30 | 0.2 | 0.2 | 500 |
| | Time/hour | 12 | 12 | 0.5 | 0.5 | 0.5 | 0.5 | 24 | 3 | 24 | 48 | 48 | 7 |
| Ref | | 73 | | 71, 68, 67 | | | | 68 | 74ᵉ | 74 | 76 | | 80 |

Cont'd

| 12 | 13 | 14ᵈ | 15ᵈ | 16ᵈ | 17 | 18 | 19ª | 20 | 21 | 22 | 23 | 24 |
|---|---|---|---|---|---|---|---|---|---|---|---|---|
| 20 | 200 | 200 | 200 | 2 | 0.2 | 1 μCi | 0.02 wt% | 2500 | 2500 | 2 | 10 | 0.2 wt% |
| 20 | 24 | 8 | 8 | 8 | 8 | 1/3 | 18 | 72 | 72 | 24 | 1 | 24 |
| 83 | 77 | 84 | | 86 | 85 | 87 | | 88 | | 90 | 89 | 103 |

Cont'd

| Compound | | 7 | 20 | 21 | 22 | 23 | 24 |
|---|---|---|---|---|---|---|---|
| Animal | Dose | 100 mg/Kg | 36 mg/Kg | 36 mg/Kg | 5 nM | 200 μM | 1 wt% |
| | Time | 48 h | 6d | 6d | 24 h | 24 h | 14d |
| Ref | | 74 ᵍ | 88 | | 90 | 89 | 103 |

ᵃinside cells
ᵇmembranes
ᶜoutside cells
ᵈGlogi
ᵉ7 (500 *** g/mL) co-assembly with Indocyanine Green (10 *** g/mL)
ᶠcell surface
ᵍ7 co-assembly with ICG(10 mg/Kg)

Besides, we rearrange the chemical structures of the precursors to directly distinguish the difference among them and easy to name them in this chapter (Scheme 4.1).

## 4.4.1 Hydrolysis of Esters

### 4.4.1.1 Phosphatase

As early as in 2004, a pioneer work finished by Yang and Xu reports the first example of enzymatic formation of supramolecular hydrogel [62]. It is a new way to turn the solution into gel using an enzyme (alkaline phosphatase) without further external stimuli. Though limited biological applications were envisioned at that time, the concept in this work has led the later-on tremendous development in this field. With proper design of the small molecules (hydrogelator precursors), it is possible to realize the hydrogelation in physiological conditions [59] and develop

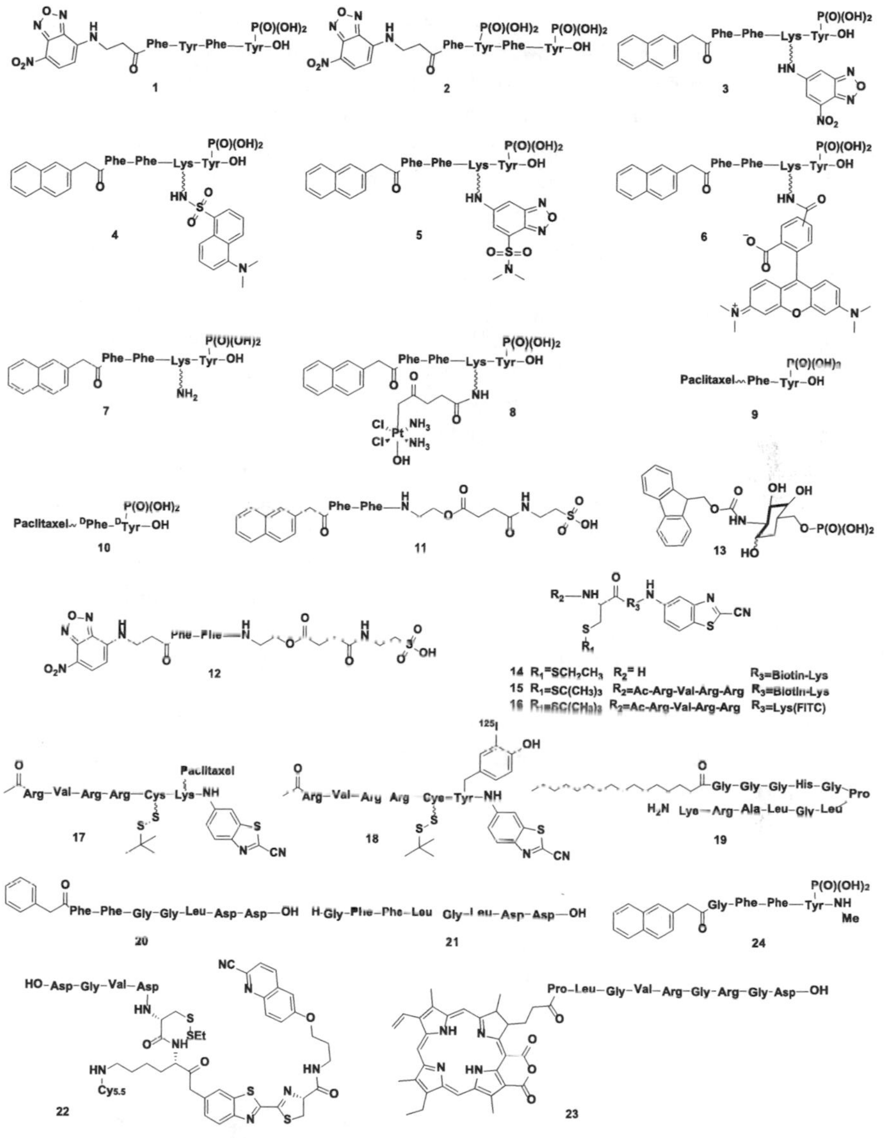

Scheme 4.1 Chemical structures of each compound numbered

applications in various directions ranging from drug controlled release, controlling cell fate, tissue engineering, and so on.

Later the Xu group reports the use of phosphatase to confer a hydrogel of paclitaxel derivative. The paclitaxel derivative is made from paclitaxel and a peptidic self-assembly motif linked by a succinic acid. Once the paclitaxel derivative is dissolved in water, phosphatase could initiate the self-assembly process and yield the formation of numerous nanofibers that result in a supramolecular hydrogel in macroscope (Fig. 4.1). Upon slow hydrolysis reaction, the hydrogel serves as a platform for controlled release of paclitaxel with adjustable release rates. Overall, this design provides a powerful method to create molecular hydrogels of clinically used therapeutics without compromising their bioactivities [66].

Then in 2012, the Xu group reported a method to image EISA of a small molecular (See Scheme 1 compound **3**) inside live cell [67] (Fig. 4.2). Aided with [31] P NMR and rheology, they demonstrated that enzyme-trigged conversion of the precursor (compound **3**) to a hydrogelator results in the formation of a hydrogel via self-assembly in vitro. Therefore, the same process is capable to perform inside living mammalian cells. The intracellular self-assembly is dependent on both the concentration of the precursor and the activity of protein tyrosine phosphatase 1B. Since the enzyme is localized on the endoplasmic reticulum (ER), it also dominates the location for the occurrence of intracellular self-assembly. The similar phenomenon is further confirmed via a co-assembly strategy to visualize the self-assembly of non-fluorescent small molecules (compound **7**) inside live cells in another work [68]. The cell fractionation experiment points out the cellular fraction containing ER triggers the fastest sol-gel transition, which implies the self-assembly occurs on ER with the highest possibility inside intact cells. Further immune co-staining shows the maximum distribution overlap between self-assemblies and ER tracker instead of lysosome or Golgi tracker. Although significant effort has been paid including correlated light and electron microscopy (CLEM), it remains difficult to identify the unambiguous fibrillar morphology of supramolecular assemblies in situ due to the intrinsic background noise from cellular components, which have similar chemical composition, the atoms of carbon, hydrogen, and oxygen. One possible way to solve this problem could be the involvement of small angle neutron scattering (SANS) which is the powerful technique investigating the internal structures of soft materials [69, 70]. To fulfill the specific aim, it requires further efforts on the correlation between SANS measurements and classical imaging data, the balance of water and heavy water to tune the scattering length density (SLD) mask the background from cell and so on, which demand too much and hard to realize.

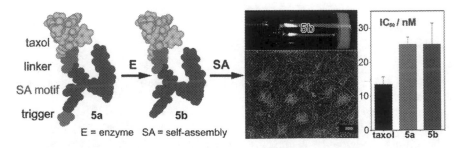

**Fig. 4.1** Schematic representation of phosphatase-triggered self-assembly of a derivative of taxol and its microstructures and the histogram of the $IC_{50}$ [66] Adapted with permission from Ref. [66]. Copyright 2009 American Chemical Society

Following the discovery of the intracellular supramolecular self-assembly, the Xu group tested a serial of different fluorophores labeled self-assembly motif against HeLa cells (Scheme 4.1, compound **3, 4, 5, 6**) [71]. Interestingly, the slight structural difference among each fluorophore has great influence on the self-assembly propensity. Besides the NBD one which can self-assemble inside cells, DBD-derived molecules could form abundant nanofibers before in contact with cell/phosphatase; Dansyl-derived molecules show certain degree of toxicity and disrupt the integrity of cell membrane while Rhodamine-derived molecules fail to form nanofibers and remain homogeneous all the time. Overall, the variant self-assembly propensity induces the drastically different distributions of self-assemblies in the cellular environment (Fig. 4.2).

It is worth mentioning that the very recent result from the Xu group demonstrated a novel way to localize supramolecular assemblies to sub-cellular organelles. The attachment of a triphenylphosphonium to the self-assembly motif acquired targeting capability to intracellular mitochondrial, which eventually can selectively kill cancer cells via disrupting the cells' powerhouse [72]. This rational design shows the potential bioactivity of intracellular self-assembly via interacting with specific sub-cellular organelles.

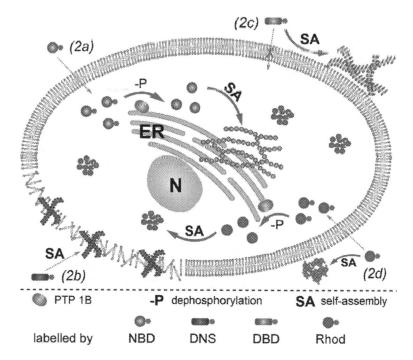

**Fig. 4.2** Illustration of the distinct spatial distribution of the small molecules in a cellular environment because of their different propensities of self-assembly before or after dephosphorylation [71] Adapted with permission from Ref. [71]. Copyright 2017 American Chemical Society. Copyright 2013 American Chemical Society

Although the kinetics of the formation of molecular assemblies is one of important features of cells, it received little attention to develop anticancer therapeutics. In 2016, the Xu group designed a serial of tetrapeptide derivatives with same sequence but a different number of phosphorous esters within the molecule. A detailed comparison showed that the variation of phosphorous esters was able to regulate the rate of supramolecular self-assembly although both the enzymatic dephosphorylation and the final hydrogelator are exactly the same (Scheme 4.1, compound **1** and **2**) [73].

Besides the cell experiments, the Chen group further applied the phosphatase-instructed self-assembly to an animal study. They adopted ICG (indocyanine green) to nanofibers formed by compound **7** to achieve co-assemblies for cancer theranostics [74]. As the first example, they form tumor-specific ICG-doped nanofibers in vivo, which is capable to manipulate the spatiotemporal distribution of ICG in mice. As a result, the prolonged retention of therapeutic agent inside tumor eventually improves the cancer theranostics.

Since the D-peptides are presumably resistant to most of enzymes, a serial of valuable studies demonstrates that the formation of the nanofibers via enzymatic dephosphorylation is independent from chirality. That means D/L enantiomers undergo the quite similar enzymatic self-assembly process. Overall, D-peptide-based EISA could achieve both bio-stability and additional desired functions simultaneously. (See Scheme 4.1 compound **8** [75], **9**, **10** [76] and **13** [77]) Very recently, the Xu group reported that the D-peptidic nanofibrils can serve as multifaceted apoptotic inducers to target cancer cells in situ. With ALP as the catalyst, D-peptidic nanofibrils which formed in situ on cancer cells presented autocrine proapoptotic ligands to their cognate receptors in a juxtacrine manner, as well as directly clustered the death receptors, which eventually activate extrinsic death signaling for selectively killing cancer cells [78]. In another study, they co-cultured a group of cancer cells and stromal cells with a D-peptidic derivative. Based on differential EISA formation of fluorescent, non-diffusive nanofibrils, the significantly higher activity of ALP on cancer cells than stromal cells was confirmed. The inherent and dynamic ALP activity was determined by drug-sensitive or drug-resistant cancer cells and even with or without hormonal stimulation [79].

### 4.4.1.2 Carboxylesterase

It is well known that the cisplatin is one of the most successful therapeutic agents for the ovarian cancers. However, the emerging drug resistance still remains a major problem in its chemotherapy treatment. To overcome such a disadvantage, the combination of cisplatin with other therapies is in an urgent need. In 2015, the Xu group firstly demonstrated that enzyme-instructed intracellular self-assembly molecular is a new approach to boost the activity of CDDP against two drug-resistant ovarian cancer cell lines. They synthesized small peptide precursors (See Scheme 4.1 compound **11**) as the substrates of carboxylesterase (CES). CES efficiently cleaved the ester bond preinstalled on the precursors and initiated the

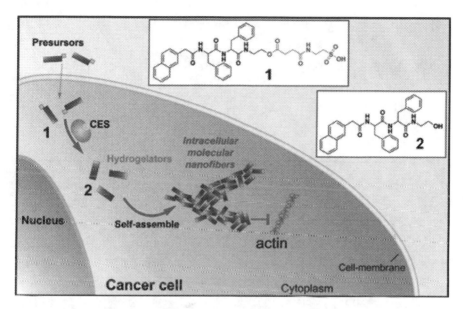

**Fig. 4.3** Enzymatic transformation of the precursor (1) (compound **11**) as a substrate of carboxylesterase (CES) to the corresponding hydrogelator (2) for intracellular self-assembly. CES efficiently cleaved the ester bond preinstalled on the compound **11** and yielded the hydrolyzed peptidic part to form nanofibers in cancer cell [80] Adapted with permission from Ref. [80]. Copyright 2015 Wiley-VCH Verlag GmbH & Co. KGaA, Weinheim

hydrolyzed peptidic part to form nanofibers in water via self-assembly. The precursors are innocuous to cells at the optimal concentrations, but they double or triple the activity of cisplatin against the drug-resistant ovarian cancer cells [80] (Fig. 4.3).

And recently, the Xu group designed peptidic precursors as the substrates of both CES and ALP to show the combination of enzyme-instructed assembly and disassembly. The precursors can turn into self-assembling molecules to form nanofibrils upon dephosphorylation by ALP, while the following CES catalyzed cleavage of the ester bond on the same molecules will result in disassembly of the nanofibrils [81]. Besides, they also illustrated a fundamentally new approach to amplify the enzymatic difference between cancer and normal cells for overcoming cancer drug resistance by establishing cytosolic EISA catalyzed by CES. The selectivity of EISA targeting cancer cells was validated in cancer and normal cell co-culture test [82].

### 4.4.1.3 Esterase

D-peptides have great applications in many areas of biology and biomedicine because of their bio-stability. However, it is ineffective for cellular uptake of D-peptides. To further explore the merits of D-peptides inside cells, it is essential to develop an effective strategy to enhance the cellular uptake of D-peptides. In 2015, the Xu group synthesized the conjugate of taurine and D-peptide which drastically

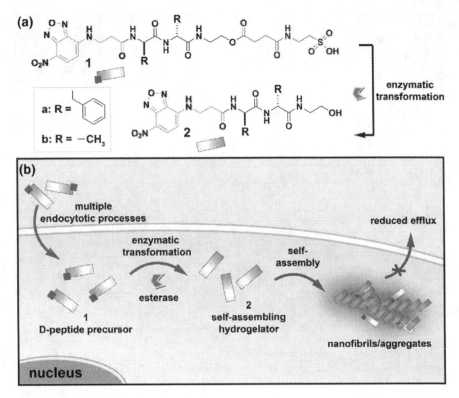

**Fig. 4.4** Molecular structures of precursors (compound 12) and the corresponding hydrogelators after enzymatic transformation. (B) Taurine conjugation boosts cellular uptake of D-peptide precursors and subsequent EISA to form nanofibers, accumulating inside cells [83] Adapted with permission from Ref. [83]. Copyright 2015 American Chemical Society

boosts the cellular uptake of small D-peptides in mammalian cells and allows intracellular esterase to trigger intracellular self-assembly of the D-peptide derivative, further enhancing their intracellular accumulation (Fig. 4.4). Using taurine for EISA is a new way to promote the uptake of bioactive molecules and generate higher-order molecular assemblies. This approach will be useful to facilitate the transport of functional D-peptides, and other bioactive molecules through the cell membrane into live cells for controlling the fate of cells [83].

## 4.4.2 Hydrolysis of Peptides

### 4.4.2.1 Furin

A controlled enzyme-triggered self-assembly in living cells was reported by Liang and the co-workers in 2010 [84]. Based on the chemical reaction between 2-cyanobenzothiazole (CBT) and D-cysteine, the authors designed a variety of thiol- or amino-protected monomers via a disulfide bond and by a peptide sequence, respectively. The protecting group of the monomers can be removed by either pH, disulfide reduction or enzymatic cleavage in vitro and then the yielded product self-assembled into different structural patterns.

Then, the authors attempted the enzyme-triggered self-assembly of monomers in live cells. The direct imaging indicated that the self-assembled signals are closely located to the furin enzyme sites–the Golgi bodies. While in control experiments where furin inhibitor was added together with monomers, the fluorescence staining pattern previously observed was absent, confirming that furin was responsible for the localized condensation of monomers at the Golgi bodies.

Based on the biocompatible condensation and subsequent self-assembly, Liang group designed a radioactive probe to self-assemble into the radioactive nanoparticles (compound **18**) under the action of intracellular glutathione (GSH) and furin in living cells [85]. These as-prepared radioactive nanoparticles could concentrate the radioactive isotopes inside cells and hamper themselves to be eliminated from the cells due to the large size and hydrophobic property. The strategy provided a new method to design smart probes for molecular imaging. Following the successful strategy, Liang Group rationally designed a taxol derivative (CBT-Taxol, compound **17**) which could condensate and self-assembled into taxol nanodrugs (Taxol-NPs) under the control of furin [86]. In vitro and in vivo studies showed that the CBT-Taxol had a 4.5 fold or 1.5-fold increase in anti-multidrug resistance effects, indicating this strategy could be used for overcoming multidrug resistance (Fig. 4.5).

### 4.4.2.2   MMP

In 2015, Maruyama and co-workers reported the enzyme-triggered molecular self-assembly of a low-molecular-weight gelator for novel anticancer applications. Its precursor (**ER-C16**, compound **19**) exhibited remarkable cytotoxicity to several cancer cell lines and low cytotoxicity to normal cells. Cancer cells secrete excessive amounts of MMP-7(matrix metalloproteinase-7), which converted the precursor into a supramolecular gelator prior to its uptake by the cells. Once the supramolecular gelators entered inside the cells, they self-assembled to form nanofibers that greatly impaired cellular function and then caused the death of the cancer cells (Fig. 4.6). Due to the unique cytotoxic mechanism, cancer cells will be unlikely to acquire resistance to the present strategy [87] (Fig. 4.6).

Besides, in 2016, the Rein V. Ulijn group also has halted tumor growth by MMP-9 triggered self-assembly of doxorubicin nanofiber depots. They demonstrated that the fibrillar depots are formed where the MMP-9 is overexpressed. And this will enhance the efficacy of doxorubicin, resulting in inhibiting of the growth of tumor in mice [88].

Due to the intimate relationships between morphology of self-assembled nanostructures and their biological performances, Wang group rationally designed a responsive   small-molecule   precursor   (compound   **23**)   that   simultaneously

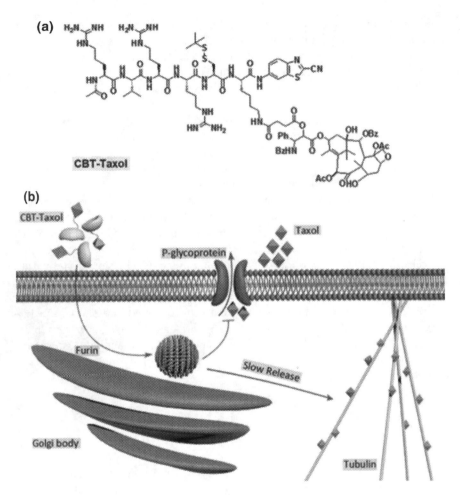

**Fig. 4.5 a** Chemical structures of CBT-Taxol. **b** Schematic illustration of intracellular furin-controlled self-assembly of Taxol-NPs for anti-MDR [86] Adapted with permission from Ref. [86]. Copyright 2009 American Chemical Society

self-assembled into nanofibers in tumor sites in 2015. The compound **23** consisted of P18, PLGVRG, and RGD. At first, the RGD can target to $\alpha_v\beta_3$ integrins, overexpressed on cancer cell membranes. Second, gelatinase, belonged to MMP-2, can cut the PLGVRD linker in tumor microenvironment. Therefore, the molecular become more hydrophobic, resulting in self-assembly of the building blocks. Eventually, the compound 23 simultaneously self-assembled into nanofibers in tumor site, and it exhibited prolonged retention time in tumor cell that directly led to an enhanced photoacoustic signal and therapeutic efficacy [89].

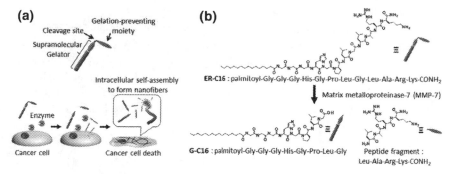

**Fig. 4.6** **a** Cancer cell death induced by molecular self-assembly of an enzyme-responsive supramolecular gelator and **b** molecular structures of compound **19** [87] Adapted with permission from Ref. [87]. Copyright 2015 American Chemical Society

### 4.4.2.3 Caspase

Arguably, bio-orthogonal click chemistry (such as Staudinger ligation, azide-alkyne, and Pictet-Spengler ligation) has already been widely applied in living cells. In these examples, after activation, small molecules can enter cells and be self-assembly. However, there are few successful examples in living animals because of the more complex environment than cultured cells. Therefore, in 2014, Rao group report a new method to direct the synthetic small molecules into nanoaggregates in living mice. They designed a fluorescent small molecule (C-SNAF, Compound **22**) and initiated its self-assembly process via caspase

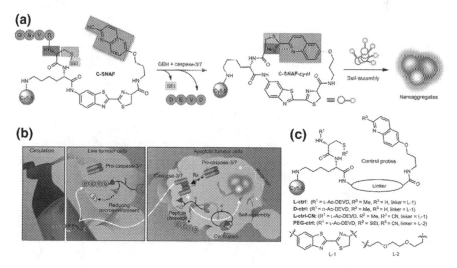

**Fig. 4.7** Illustration of the mechanism of in vivo imaging by C-SNAF of caspase-3/7 activity in human tumor xenograft mouse models [90] Adapted with permission from Ref. [90]. Copyright 2014 Nature Publishing Group

activation (Fig. 4.7). The compound 22 comprised of two parts. One part was
D-cysteine and 2-cyano-6-hydroxyquinoline moieties linked to an amino luciferin
scaffold. The rest consisted of an L-DEVD (Asp-Gly-Val-Asp) capping sequence
and a disulfide bond. The self-assembly process required a two-step activation,
which was caspase-mediated cleavage and intracellular thiol-mediated reduction.
The strategy combined the advantages offered by small molecules with those of
nanomaterials and should find widespread use for non-invasive imaging of enzyme
activity in vivo [90].

## 4.5   Vaccine Adjuvants

Vaccines play a very important role in our daily life and have been widely applied
in medical fields by preventing the body against or treat the diseases through
balancing the immune response to be immunity or silence [91]. With the development of vaccination, it probably serves as a medical intervention. The example is
the first therapeutic cancer vaccine, which was licensed in 2010 [92]. Due to the
complexity of human immune system and our lack of understanding of it, vaccine
design still is challenging. Therefore, DNA vaccines may be an ideal method,
which can generate long-term humoral and cellular immune responses [93, 94]. As
far, there have developed many delivery systems to improve the efficiency of
delivery DNA into mammalian cells, such as liposomes, [95] nanoparticles [96],
and polymers [97, 98]. Despite of the advantages of these systems, there are still
some disadvantages to overcome, including high toxicity, [99] low amount of
antigen loading [100] and reduce the bioactivity of DNA in modification [101,
102]. Therefore, EISA may be an ideal system to deliver DNA and the
self-assemblies base on EISA to produce vaccines have been studied and designed.

HIV infection is incurable so far, and developing efficient vaccines is urgent.
Yang and co-workers designed a nanovector in 2014, which can condense DNA to
result in strong immune responses against HIV [103]. This nanovector composed of
peptide-based nanofibrous hydrogel can strongly induce both tumoral and cellular
immune response of HIV Env DNA to a balanced level, which was rarely reported
in previous studies (Fig. 4.8). The nanovector shows good biocompatibility both
in vitro and in vivo, and the well-defined nanofibrous structure is significantly
important for the enhanced immune responses. These nanofibrous hydrogels are
potentially applicable to the development of efficacious HIV DNA vaccines.

And subsequently, Yang's group reported that in suit-formed peptidic nanofibers
facilitate the induction of multiple crucial immunities against HIV DNA vaccine,
including multi-functional T cell response, broad IgG subclasses response, and V1/
V2-specific IgG response [104]. And then they reported the first co-assembly of
peptide and protein (ovalbumin, OVA) upon alkaline phosphatase (ALP) catalysis
[105]. The results show an obvious increase of the IgG production when compared
to the clinically used alum adjuvant and thus have big potential for application in
immunotherapy against different diseases as protein vaccines.

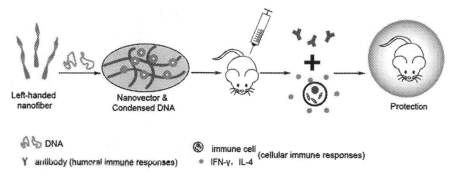

Left-handed nanofiber    Nanovector & Condensed DNA    Protection

𝑁 DNA    ⊛ immune cell (cellular immune responses)
Y antibody (humoral immune responses)    ⊛ IFN-γ, IL-4

**Fig. 4.8** Process of peptide-based nanofibrous hydrogel for enhancing immune responses of HIV DNA vaccines [103] Adapted with permission from Ref. [103]. Copyright 2014 American Chemical Society

The studies of using the peptide-based vaccines as physical carriers have showed great potential applications. And another approach was also used for designing such peptide-based vaccines by the previous conjunction of assembling peptide and model antigens or proteins. Collier and co-workers deigned peptide-based assembling supramolecular containing strong epitopes and demonstrated that the nanofiber induced strong antibody responses in mice [106–108]. MUCI proteins have a lot of tandem repeats, bearing tumor-associated carbohydrate antigens, which is the key problem to develop vaccines for epithelial tumors. To overcome the low immunogenicity of the short MCUI peptide, Li and co-workers designed the self-adjuvanting vaccine with self-assembly domains, which combined some vaccine candidates with a self-assembly peptide sequence. These vaccines can induce antibodies that recognized human breast tumor cells without additional adjuvant. It is reported that these vaccines can act through a T cell independent pathway and may be associated with the activity of cytotoxic T cells [109].

## 4.6 Outlook

Despite the rapid outcomes about enzyme-activated theranostics under the EISA strategy, it remains, as well as the whole precision nanomedicine field, lack of real applications for improving the overall benefits for patients. Thus, there is still plenty of room to be perfected. As of EISA, we would raise at least three aspects here.

(1) Elucidate the micromorphology of EISA in biological milieu. Various microscopic techniques, such as AFM, SEM, TEM, and so on are working on very well to study the micromorphology of EISA in vitro. However, because of the complexity of the cell itself, it remains impossible to directly capture a vivid snapshot of EISA with nanometer resolution in biological milieu. The attempts on heavy atom labeling (such as Iodine or certain metal element) may be an

option to allow the EISA visible under environmental electron microscope. The advancement in super-resolution fluorescent microscope may be another potential technique to conquer this task since multi-color fluorescent probe labeling is feasible and practical.

(2) Involve enzymes other than hydrolase, or even metabolites to instruct self-assembly. Although there have been plenty of strategies to generate EISA, however, all the enzymes involved in EISA monotonically belong to hydrolase —one of six type of enzymes based on catalyzed chemical reactions in the top-level enzyme classification. In comparison with hydrolase, some other enzymes may be of even greater importance in cancer therapy. For example, kinase is one of the most attractive targets in anticancer practice. This concept has drawn numerous money for the discovery of adequate kinase inhibitor which may eventually suppress the growth of cancer. However, more recent studies are likely to disclose kinase as "untargetable." Thus, if we can develop a system which uses kinase to instruct the self-assembly in vivo, it will be of great value helping the progress in kinase-oriented cancer therapy.

In addition to novel ways to instruct self-assembly, the precise localization remains of great interest which is critical to eliminate false-positive/negative results. Current EISA usually response to single stimuli such as abnormal enzyme expression and its activity. The involvement of two enzymes to instruct EISA process makes this strategy more delicate and potentially targetable to a broader range of cancers [49]. Since there are plenty of known hallmarks of cancers, a well-designed molecule which can respond to those stimulus and perform multi-step tandem transformation before self-assemble will definitely enhance the targeting capability and precision of EISA.

(3) Exploit the interaction between EISA and sub-cellular organelles. Most of the enzymes are associated with certain sub-cellular organelles, e.g., the alkaline phosphatase on the cell membrane, protein tyrosine phosphatase 1B on the endoplasmic reticulum, furin in the Golgi. As the formation of EISA is initiated by enzymatic reactions, the location of correlated enzyme dominates the distribution of EISA. We have realized this interesting phenomenon, but there is little discussion on the interaction between EISA and "host" sub-cellular organelles. The exploration on this point would be useful in controlling the cell fate by regulating the activity of its sub-cellular organelles.

(4) Incorporation with bio-orthogonal chemistry. EISA provides an excellent strategy for the construction of nanomaterial inside biological environment with targeting capability. But we notice that the required CAC is somehow deviated from the dosage of traditional drug molecules. If we covalently load drug molecule on self-assembly motif (that is 1:1 molar ratio), there must be an imbalance between severe overdose issue and initiation of self-assembly. To solve this inevitable dilemma, we think the incorporation of bio-orthogonal reaction would be a realistic solution. By linking a bio-orthogonal reaction handle (azide, alkyne, tetrazine, and so on) to a self-assembly motif, the in situ formed EISA material composed of those as-prepared functionalize molecule

will spontaneously deliver and localize bio-orthogonal reaction handles. A following therapeutic agent modified with the counter handle will efficiently target EISA in the desired location. We envision this pathway owns several advantages including the on/off target ratio, the adjustable dosing, and even the application of toxin which is too toxic by itself.

# References

1. Siegel JS (1996) Supramolecular chemistry. Concepts and perspectives - Lehn, JM. Science 271(5251):949–949
2. Whitesides GM, Grzybowski B (2002) Self-assembly at all scales. Science 295(5564):2418–2421
3. Marler JJ, Upton J, Langer R, Vacanti JP (1998) Transplantation of cells in matrices for tissue regeneration. Adv Drug Deliv Rev 33(1–2):165–182
4. Lee KY, Mooney DJ (2001) Hydrogels for tissue engineering. Chem Rev 101(7):1869–1879
5. Andrianov AK, Cohen S, Visscher KB, Payne LG, Allcock HR, Langer R (1993) Controlled-release using ionotropic polyphosphazene hydrogels. J Control Release 27(1):69–77
6. Xing BG, Yu CW, Chow KH, Ho PL, Fu DG, Xu B (2002) Hydrophobic interaction and hydrogen bonding cooperatively confer a vancomycin hydrogel: a potential candidate for biomaterials. J Am Chem Soc 124(50):14846–14847
7. Terech P, Ostuni E, Weiss RG (1996) Structural study of cholesteryl anthraquinone-2 carboxylate (CAQ) physical organogels by neutron and X-ray small angle scattering. J Phys Chem 100(9):3759–3766
8. Wang R, Geiger C, Chen LH, Swanson B, Whitten DG (2000) Direct observation of sol-gel conversion: the role of the solvent in organogel formation. J Am Chem Soc 122(10):2399–2400
9. Terech P, Weiss RG (1997) Low molecular mass gelators of organic liquids and the properties of their gels. Chem Rev 97(8):3133–3159
10. Piepenbrock MOM, Lloyd GO, Clarke N, Steed JW (2010) Metal- and anion-binding supramolecular gels. Chem Rev 110(4):1960–2004
11. Yan XH, Zhu PL, Li JB (2010) Self-assembly and application of diphenylalanine-based nanostructures. Chem Soc Rev 39(6):1877–1890
12. Xu X, Jha AK, Harrington DA, Farach-Carson MC, Jia XQ (2012) Hyaluronic acid-based hydrogels: from a natural polysaccharide to complex networks. Soft Mat 8(12):3280–3294
13. Ajayaghosh A, Praveen VK, Vijayakumar C (2008) Organogels as scaffolds for excitation energy transfer and light harvesting. Chem Soc Rev 37(1):109–122
14. Diaz DD, Kuhbeck D, Koopmans RJ (2011) Stimuli-responsive gels as reaction vessels and reusable catalysts. Chem Soc Rev 40(1):427–448
15. Simonsen AC, Rubahn HG (2002) Local spectroscopy of individual hexaphenyl nanofibers. Nano Lett 2(12):1379–1382
16. Maggini L, Bonifazi D (2012) Hierarchised luminescent organic architectures: design, synthesis, self-assembly, self-organisation and functions. Chem Soc Rev 41(1):211–241
17. Lin YC, Weiss RG (1989) Evidence for random parallel and anti-parallel packing between neighboring cholesteryl 4-(2-Anthryloxy)Butyrate (Cab) molecules in the cholesteric liquid-crystalline phase identification of the 4 photodimers from cab. Liq Cryst 4(4):367–384
18. Brotin T, Utermohlen R, Fages F, Bouaslaurent H, Desvergne JP (1991) A novel small molecular luminescent gelling agent for alcohols. J Chem Soc Chem Comm 6:416–418

19. Oner D, McCarthy TJ (2000) Ultrahydrophobic surfaces. Effects of topography length scales on wettability. Langmuir 16(20):7777–7782

20. Maitra U, Kumar PV, Chandra N, D'Souza LJ, Prasanna MD, Raju AR (1999) First donor-acceptor interaction promoted gelation of organic fluids. Chem Commun 7:595–596

21. Mukkamala R, Weiss RG (1996) Physical gelation of organic fluids by anthraquinone-steroid-based molecules. Structural features influencing the properties of gels. Langmuir 12(6):1474–1482

22. Furman I, Weiss RG (1993) Liquid-crystalline solvents as mechanistic probes. 49. Factors influencing the formation of thermally reversible gels comprised of cholesteryl 4-(2-Anthryloxy)Butanoate in hexadecane, 1-octanol, or their mixtures. Langmuir 9 (8):2084–2088

23. Gortner RA, Hoffman WF (1921) An interesting colloid gel. J Am Chem Soc 43:2199–2202

24. Menger FM, Caran KL (2000) Anatomy of a gel. Amino acid derivatives that rigidify water at submillimolar concentrations. J Am Chem Soc 122(47):11679–11691

25. Vonderviszt F, Sonoyama M, Tasumi M, Namba K (1992) Conformational adaptability of the terminal regions of flagellin. Biophys J 63(6):1672–1677

26. Malkoch M, Vestberg R, Gupta N, Mespouille L, Dubois P, Mason AF, Hedrick JL, Liao Q, Frank CW, Kingsbury K, Hawker CJ (2006) Synthesis of well-defined hydrogel networks using Click chemistry. Chem Commun 0(26):2774–2776

27. Heller A (1990) Electrical wiring of redox enzymes. Acc Chem Res 23(5):128–134

28. Mao F, Mano N, Heller A (2003) Long tethers binding redox centers to polymer backbones enhance electron transport in enzyme "wiring" hydrogels. J Am Chem Soc 125(16): 4951–4957

29. Elbert DL, Pratt AB, Lutolf MP, Halstenberg S, Hubbell JA (2001) Protein delivery from materials formed by self-selective conjugate addition reactions. J Control Release 76(1–2):11–25

30. Zhao F, Gao YA, Shi JF, Browdy HM, Xu B (2011) Novel Anisotropic supramolecular hydrogel with high stability over a wide ph range. Langmuir 27(4):1510–1512

31. Wathier M, Johnson CS, Kim T, Grinstaff MW (2006) Hydrogels formed by multiple peptide ligation reactions to fasten corneal transplants. Bioconjug Chem 17(4):873–876

32. Ikeda M, Ochi R, Kurita Y, Pochan DJ, Hamachi I (2012) Heat-Induced morphological transformation of supramolecular nanostructures by Retro-Diels-Alder reaction. Chem Eur J 18(41):13091–13096

33. Printz C (2015) Precision medicine initiative boosts funding for NCI efforts proposal would help broaden availability of targeted therapies. Cancer 121(19):3369–3370

34. Erdogan H, Yilmaz M, Babur E, Duman M, Aydin HM, Demirel G (2016) Fabrication of plasmonic nanorod-embedded dipeptide microspheres via the freeze-quenching method for near-infrared laser-triggered drug-delivery applications. Biomacromol 17(5):1788–1794

35. Karimi M, Ghasemi A, Zangabad PS, Rahighi R, Basri SMM, Mirshekari H, Amiri M, Pishabad ZS, Aslani A, Bozorgomid M, Ghosh D, Beyzavi A, Vaseghi A, Aref AR, Haghani L, Bahrami S, Hamblin MR (2016) Smart micro/nanoparticles in stimulus-responsive drug/gene delivery systems. Chem Soc Rev 45(5):1457–1501

36. Whitlow J, Pacelli S, Paul A (2016) Polymeric nanohybrids as a new class of therapeutic biotransporters. Macromol Chem Phys 217(11):1245–1259

37. Alcantara D, Lopez S, Garcia-Martin ML, Pozo D (2016) Iron oxide nanoparticles as magnetic relaxation switching (MRSw) sensors: current applications in nanomedicine. Nanomed Nanotechnol 12(5):1253–1262

38. Ohta S, Glancy D, Chan WCW (2016) Nanomaterials DNA-controlled dynamic colloidal nanoparticle systems for mediating cellular interaction. Science 351(6275):841–845

39. Sindhwani S, Syed AM, Wilhelm S, Glancy DR, Chen YY, Dobosz M, Chan WCW (2016) Three-dimensional optical mapping of nanoparticle distribution in intact tissues. ACS Nano 10(5):5468–5478

40. Ferrari M (2005) Cancer nanotechnology: Opportunities and challenges. Nat Rev Cancer 5(3):161–171

41. Peer D, Karp JM, Hong S, FaroKHzad OC, Margalit R, Langer R (2007) Nanocarriers as an emerging platform for cancer therapy. Nat Nanotechnol 2(12):751–760
42. Kanapathipillai M, Brock A, Ingber DE (2014) Nanoparticle targeting of anti-cancer drugs that alter intracellular signaling or influence the tumor microenvironment. Adv Drug Deliver Rev 79–80:107–118
43. Howes PD, Chandrawati R, Stevens MM (2014) Colloidal nanoparticles as advanced biological sensors. Science 346(6205):1247390–1247400
44. Petersen GH, Alzghari SK, Chee W, Sankari SS, La-Beck NM (2016) Meta-analysis of clinical and preclinical studies comparing the anticancer efficacy of liposomal versus conventional non-liposomal doxorubicin. J Control Release 232:255–264
45. Wilhelm S, Tavares AJ, Dai Q, Ohta S, Audet J, Dvorak HF, Chan WCW (2016) Analysis of nanoparticle delivery to tumours. Nat Rev Mater 1:16014
46. Versluis F, van Esch JH, Eelkema R (2016) Synthetic self-assembled materials in biological environments. Adv Mater 28(23):4576–4592
47. Li Y, Cui R, Zhang P, Chen B-B, Tian Z-Q, Li L, Hu B, Pang D-W, Xie Z-X (2013) Mechanism-oriented controllability of intracellular quantum dots formation: the role of glutathione metabolic pathway. ACS Nano 7(3):2240–2248
48. Zhou J, Xu B (2015) Enzyme-instructed self-assembly: a multistep process for potential cancer therapy. Bioconjug Chem 26(6):987–999
49. Zhou J, Li J, Du XW, Xu B (2017) Supramolecular biofunctional materials. Biomaterials 129:1–27
50. Zhou J, Du X, Xu B (2015) Prion-like nanofibrils of small molecules (PriSM): a new frontier at the intersection of supramolecular chemistry and cell biology. Prion 9(2):110–118
51. Du X, Zhou J, Shi J, Xu B (2015) Supramolecular hydrogelators and hydrogels: from soft matter to molecular biomaterials. Chem Rev 115(24):13165–13307
52. Boekhoven J, Poolman JM, Maity C, Li F, van der Mee L, Minkenberg CB, Mendes E, van Esch JH, Eelkema R (2013) Catalytic control over supramolecular gel formation. Nat Chem 5(5):433–437
53. Boekhoven J, Hendriksen WE, Koper GJM, Eelkema R, van Esch JH (2015) Transient assembly of active materials fueled by a chemical reaction. Science 349(6252):1075–1079
54. Onogi S, Shigemitsu H, Yoshii T, Tanida T, Ikeda M, Kubota R, Hamachi I (2016) In situ real-time imaging of self-sorted supramolecular nanofibres. Nat Chem 8(8):743–752
55. Cui HG, Cheetham AG, Pashuck ET, Stupp SI (2014) Amino acid sequence in constitutionally isomeric tetrapeptide amphiphiles dictates architecture of one-dimensional nanostructures. J Am Chem Soc 136(35):12461–12468
56. Aida T, Meijer EW, Stupp SI (2012) Functional Supramolecular polymers. Science 335 (6070):813–817
57. Estroff LA, Hamilton AD (2004) Water gelation by small organic molecules. Chem Rev 104(3):1201–1217
58. Korevaar PA, Grenier C, Markvoort AJ, Schenning APHJ, de Greef TFA, Meijer EW (2013) Model-driven optimization of multicomponent self-assembly processes. P Natl Acad Sci 110(43):17205–17210
59. Yang Z, Liang G, Xu B (2008) Enzymatic hydrogelation of small molecules. Acc Chem Res 41(2):315–326
60. Fletcher DA, Mullins D (2010) Cell mechanics and the cytoskeleton. Nature 463(7280):485–492
61. Pollard TD, Cooper JA (2009) Actin, a central player in cell shape and movement. Science 326(5957):1208–1212
62. Yang ZM, Gu HW, Fu DG, Gao P, Lam JK, Xu B (2004) Enzymatic formation of supramolecular hydrogels. Adv Mater 16(16):1440–1444
63. Yang ZM, Liang GL, Wang L, Xu B (2006) Using a kinase/phosphatase switch to regulate a supramolecular hydrogel and forming the supramoleclar hydrogel in vivo. J Am Chem Soc 128(9):3038–3043

64. Wang H, Feng Z, Wu D, Fritzsching KJ, Rigney M, Zhou J, Jiang Y, Schmidt-Rohr K, Xu B (2016) Enzyme-regulated supramolecular assemblies of cholesterol conjugates against drug-resistant ovarian cancer cells. J Am Chem Soc 138(34):10758–10761
65. Zhou J, Du XW, Yamagata N, Xu B (2016) Enzyme-instructed self-assembly of small D-peptides as a multiple-step process for selectively killing cancer cells. J Am Chem Soc 138(11):3813–3823
66. Gao Y, Kuang Y, Guo ZF, Guo ZH, Krauss IJ, Xu B (2009) Enzyme-instructed molecular self-assembly confers nanofibers and a supramolecular hydrogel of taxol derivative. J Am Chem Soc 131(38):13576–13577
67. Gao Y, Shi JF, Yuan D, Xu B (2012) Imaging enzyme-triggered self-assembly of small molecules inside live cells. Nat Commun 3 (1033)
68. Gao Y, Berciu C, Kuang Y, Shi JF, Nicastro D, Xu B (2013) Probing nanoscale self-assembly of nonfluorescent small molecules inside live mammalian cells. ACS Nano 7 (10):9055–9063
69. Gao Y, Nieuwendaal R, Dimitriadis E, Hammouda B, Douglas J, Xu B, Horkay F (2016) Supramolecular self-assembly of a model hydrogelator: characterization of fiber formation and morphology. Gels 2(4):27
70. Hule RA, Nagarkar RP, Hammouda B, Schneider JP, Pochan DJ (2009) Dependence of self-assembled peptide hydrogel network structure on local fibril nanostructure. Macromolecules 42(18):7137–7145
71. Gao Y, Kuang Y, Du XW, Zhou J, Chandran P, Horkay F, Xu B (2013) Imaging self-assembly dependent spatial distribution of small molecules in a cellular environment. Langmuir 29(49):15191–15200
72. Wang HM, Feng ZQQ, Wang YZ, Zhou R, Yang ZM, Xu B (2016) Integrating enzymatic self-assembly and mitochondria targeting for selectively killing cancer cells without acquired drug resistance. J Am Chem Soc 138(49):16046–16055
73. Zhou J, Du XW, Xu B (2016) Regulating the rate of molecular self-assembly for targeting cancer cells. Angew Chem Int Ed 55(19):5770–5775
74. Huang P, Gao Y, Lin J, Hu H, Liao HS, Yan XF, Tang YX, Jin A, Song JB, Niu G, Zhang GF, Horkay F, Chen XY (2015) Tumor-specific formation of enzyme-instructed supramolecular self-assemblies as cancer theranostics. ACS Nano 9(10):9517–9527
75. Liu H, Li YL, Lyu ZL, Wan YB, Li XH, Chen HB, Chen H, Li XM (2014) Enzyme-triggered supramolecular self-assembly of platinum prodrug with enhanced tumor-selective accumulation and reduced systemic toxicity. J Mater Chem B 2(47):8303–8309
76. Huang AQ, Ou CW, Cai YB, Wang ZY, Li HK, Yang ZM, Chen MS (2016) In situ enzymatic formation of supramolecular nanofibers for efficiently killing cancer cells. RSC Adv 6(39):32519–32522
77. Pires RA, Abul-Haija YM, Costa DS, Novoa-Carballal R, Reis RL, Ulijn RV, Pashkuleva I (2015) Controlling cancer cell fate using localized biocatalytic self-assembly of an aromatic carbohydrate amphiphile. J Am Chem Soc 137(2):576–579
78. Du X, Zhou J, Wang H, Shi J, Kuang Y, Zeng W, Yang Z, Xu B (2017) In situ generated D-peptidic nanofibrils as multifaceted apoptotic inducers to target cancer cells. Cell Death Dis 8(2):e2614
79. Zhou J, Du X, Berciu C, He H, Shi J, Nicastro D, Xu B (2016) Enzyme-instructed self-assembly for spatiotemporal profiling of the activities of alkaline phosphatases on live cells. Chemicals 1(2):246–263
80. Li J, Kuang Y, Shi JF, Zhou J, Medina JE, Zhou R, Yuan D, Yang CH, Wang HM, Yang ZM, Liu JF, Dinulescu DM, Xu B (2015) Enzyme-instructed intracellular molecular self-assembly to boost activity of cisplatin against drug-resistant ovarian cancer cells. Angew Chem Int Ed 54(45):13307–13311
81. Feng Z, Wang H, Zhou R, Li J, Xu B (2017) Enzyme-instructed assembly and disassembly processes for targeting downregulation in cancer cells. J Am Chem Soc 139(11):3950–3953

82. Li J, Shi JF, Medina JE, Zhou J, Du XW, Wang HM, Yang CH, Liu JF, Yang ZM, Dinulescu DM, Xu B (2017) Selectively inducing cancer cell death by intracellular enzyme-instructed self-assembly (EISA) of dipeptide derivatives. Adv. Healthcare Mater 6(15):1601400–1601408

83. Zhou J, Du XW, Li J, Yamagata N, Xu B (2015) Taurine boosts cellular uptake of small d-peptides for enzyme-instructed intracellular molecular self-assembly. J Am Chem Soc 137 (32):10040–10043

84. Liang GL, Ren HJ, Rao JH (2010) A biocompatible condensation reaction for controlled assembly of nanostructures in living cells. Nat Chem 2(1):54–60

85. Miao QQ, Bai XY, Shen YY, Mei B, Gao JH, Li L, Liang GL (2012) Intracellular self-assembly of nanoparticles for enhancing cell uptake. Chem Commun 48(78):9738–9740

86. Yuan Y, Wang L, Du W, Ding ZL, Zhang J, Han T, An LN, Zhang HF, Liang GL (2015) Intracellular self-assembly of taxol nanoparticles for overcoming multidrug resistance. Angew Chem Int Ed 54(33):9700–9704

87. Tanaka A, Fukuoka Y, Morimoto Y, Honjo T, Koda D, Goto M, Maruyama T (2015) Cancer cell death induced by the intracellular self-assembly of an enzyme-responsive supramolecular gelator. J Am Chem Soc 137(2):770–775

88. Kalafatovic D, Nobis M, Son JY, Anderson KI, Ulijn RV (2016) MMP-9 triggered self-assembly of doxorubicin nanofiber depots halts tumor growth. Biomaterials 98:192–202

89. Zhang D, Qi GB, Zhao YX, Qiao SL, Yang C, Wang H (2015) In Situ Formation of Nanofibers from Purpurin18-peptide conjugates and the assembly induced retention effect in tumor sites. Adv Mater 27(40):6125–6130

90. Ye DJ, Shuhendler AJ, Cui LN, Tong L, Tee SS, Tikhomirov G, Felsher DW, Rao JH (2014) Bioorthogonal cyclization-mediated in situ self-assembly of small-molecule probes for imaging caspase activity in vivo. Nat Chem 6(6):519–526

91. Irvine DJ, Swartz MA, Szeto GL (2013) Engineering synthetic vaccines using cues from natural immunity. Nat Mater 12(11):978–990

92. Mellman I, Coukos G, Dranoff G (2011) Cancer immunotherapy comes of age. Nature 480 (7378):480–489

93. Wen J, Yang Y, Zhao GY, Tong S, Yu H, Jin X, Du LY, Jiang SB, Kou ZH, Zhou YS (2012) Salmonella typhi Ty21a bacterial ghost vector augments HIV-1 gp140 DNA vaccine-induced peripheral and mucosal antibody responses via TLR4 pathway. Vaccine 30 (39):5733–5739

94. Hansen SG, Ford JC, Lewis MS, Ventura AB, Hughes CM, Coyne-Johnson L, Whizin N, Oswald K, Shoemaker R, Swanson T, Legasse AW, Chiuchiolo MJ, Parks CL, Axthelm MK, Nelson JA, Jarvis MA, Piatak M, Lifson JD, Picker LJ (2011) Profound early control of highly pathogenic SIV by an effector-memory T cell vaccine. Nature 473 (7348):523–527

95. Greenland JR, Letvin NL (2007) Chemical adjuvants for plasmid DNA vaccines. Vaccine 25 (19):3731–3741

96. Jiang L, Qian F, He X, Wang F, Ren D, He Y, Li K, Sun S, Yin C (2007) Novel chitosan derivative nanoparticles enhance the immunogenicity of a DNA vaccine encoding hepatitis B virus core antigen in mice. J Gene Med 9(4):253–264

97. Qiao Y, Huang Y, Qiu C, Yue XY, Deng LD, Wan YM, Xing JF, Zhang CY, Yuan SH, Dong AJ, Xu JQ (2010) The use of PEGylated poly [2-(N, N-dimethylamino) ethyl methacrylate] as a mucosal DNA delivery vector and the activation of innate immunity and improvement of HIV-1-specific immune responses. Biomaterials 31(1):115–123

98. Minigo G, Scholzen A, Tang CK, Hanley JC, Kalkanidis M, Pietersz GA, Apostolopoulos V, Plebanski M (2007) Poly-l-lysine-coated nanoparticles: A potent delivery system to enhance DNA vaccine efficacy. Vaccine 25(7):1316–1327

99. Fischer D, Li YX, Ahlemeyer B, Krieglstein J, Kissel T (2003) In vitro cytotoxicity testing of polycations: influence of polymer structure on cell viability and hemolysis. Biomaterials 24(7):1121–1131

100. Malmsten M (2006) Soft drug delivery systems. Soft Mat 2(9):760–769

101. Walter E, Moelling K, Pavlovic J, Merkle HP (1999) Microencapsulation of DNA using poly(DL-lactide-co-glycolide): stability issues and release characteristics. J Control Release 61(3):361–374
102. Ando S, Putnam D, Pack DW, Langer R (1999) PLGA microspheres containing plasmid DNA: Preservation of supercoiled DNA via cryopreparation and carbohydrate stabilization. J Pharm 88(1):126–130
103. Tian Y, Wang HM, Liu Y, Mao LN, Chen WW, Zhu ZN, Liu WW, Zheng WF, Zhao YY, Kong DL, Yang ZM, Zhang W, Shao YM, Jiang XY (2014) A Peptide-based nanofibrous hydrogel as a promising dna nanovector for optimizing the efficacy of HIV vaccine. Nano Lett 14(3):1439–1445
104. Liu Y, Wang HM, Li D, Tian Y, Liu WW, Zhang LM, Zheng WS, Hao YL, Liu JD, Yang ZM, Shao YM, Jiang XY (2016) In situ formation of peptidic nanofibers can fundamentally optimize the quality of immune responses against HIV vaccine. Nanoscale Horiz. 1(2):135–143
105. Wang HM, Luo Z, Wang YCZ, He T, Yang CB, Ren CH, Ma LS, Gong CY, Li XY, Yang ZM (2016) Enzyme-catalyzed formation of supramolecular hydrogels as promising vaccine adjuvants. Adv Funct Mater 26(11):1822–1829
106. Wen Y, Waltman A, Han H, Collier JH (2016) Switching the immunogenicity of peptide assemblies using surface properties. ACS Nano 10(10):9274–9286
107. Pompano RR, Chen J, Verbus EA, Han H, Fridman A, McNeely T, Collier JH, Chong AS (2014) Titrating T-Cell epitopes within self-assembled vaccines optimizes CD4 + helper T cell and antibody outputs. Adv Healthcare Mater 3(11):1898–1908
108. Rudra JS, Tian YF, Jung JP, Collier JH (2010) A self-assembling peptide acting as an immune adjuvant. P Natl Acad Sci 107(2):622–627
109. Huang ZH, Shi L, Ma JW, Sun ZY, Cai H, Chen YX, Zhao YF, Li YM (2012) A totally synthetic, self-assembling, adjuvant-free MUC1 glycopeptide vaccine for cancer therapy. J Am Chem Soc 134(21):8730–8733

# Chapter 5
# Acid- and Redox-Responsive Smart Polymeric Nanomaterials for Controlled Drug Delivery

Zeng-Ying Qiao and Yu-Juan Gao

**Abstract** In cancer therapy, the lack of selectivity of anticancer drug remains a major challenge for chemotherapy. Polymeric nanomaterials have enabled more effective drug design and development through tumor passive and active targeted drug delivery strategies. Considering the possible inefficient drug release at the tumor site, stimuli-responsive chemical bonds were introduced into the polymeric nanocarriers for controlled drug release and enhanced therapeutic efficacy toward tumor. Numerous external and internal stimuli were applied to trigger the chemical structure change of polymeric nanovehicles, causing microstructural rearrangement, hydrophilic/hydrophobic balance transformation, or disassembly into single polymer chains. Characteristic changes in tumor site such as pH ($\sim$pH 6.8 in tumor microenvironment and $\sim$pH 5–6 in endosomes/lysosomes) or reductive/oxidative gradients in cytoplasm/mitochondria have been investigated extensively as internal stimuli for triggering drug release from polymeric nanocarriers. Therefore, it is important to develop acid- and redox-responsive polymeric nanomaterials as potential anticancer nanodrugs. In this review, we summarize recent progress in the preparation of these responsive polymers and their applications in anticancer drug delivery. We will focus on chemical structure, stimuli-responsive bond, self-assembly property, and drug release behavior of polymeric nanomaterials, which may offer a guide for optimal design of smart targeted nanodrugs in clinical cancer treatment.

Z.-Y. Qiao (✉) · Y.-J. Gao
CAS Center for Excellence in Nanoscience, CAS Key Laboratory for Biomedical Effects
of Nanomaterials and Nanosafety, National Center for Nanoscience and Technology
(NCNST), No. 11 Beiyitiao, Zhongguancun, Beijing, China
e-mail: qiaozy@nanoctr.cn

Y.-J. Gao
e-mail: gaoyj@nanoctr.cn

© Springer Nature Singapore Pte Ltd. 2018
H. Wang and L. L. Li (eds.), *In Vivo Self-Assembly Nanotechnology for Biomedical
Applications*, Nanomedicine and Nanotoxicology,
https://doi.org/10.1007/978-981-10-6913-0_5

## 5.1 Introduction

Cancer has been one of the leading causes of death in the world, and the incidence of cancer is dramatically increasing in recent years. A group of agents with anticancer activity that induce cell apoptosis have been a main focus of research. Advances in chemotherapy have resulted in the development of various successful cytotoxic drugs such as doxorubicin (DOX) and paclitaxel (PTX) [1]. However, there are still some major challenges for traditional chemotherapy, such as the lack of selectivity toward cells, resulting in potentially life-threatening systemic side effects [2, 3]. This emphasizes the importance of localized drug delivery, defined as a method of delivering a biologically active molecule to a particular site in the biological system where its entire pharmacological effect is desired. Therefore, drug delivery systems (DDSs) that deliver anticancer drugs selectively to the tumor sites are being explored extensively. Nanomaterials can be valuable therapeutic options to solve this problem due to their ability to selectively target the pathological area [4–11].

Nanocarriers are able to physically encapsulate or chemically conjugate a therapeutic molecule and achieve targeted delivery of their payload to solid tumors via the enhanced permeability and retention (EPR) effect (Fig. 5.1) [12–14]. In general, small molecules can diffuse into normal and tumor tissues freely through the endothelia cell layer of blood capillaries. However, nanoscaled materials cannot pass through the capillary walls of normal tissue. Compared with normal tissues, the capillaries in the tumor tissue are fenestrated and leaky so that nanoparticles can retain and subsequently accumulate in solid tumor, which is called passive targeting of nanomaterials. In order to achieve passive targeting, the nanomaterials need to be circulated in the bloodstream for sufficient time. Therefore, the ability to avoid

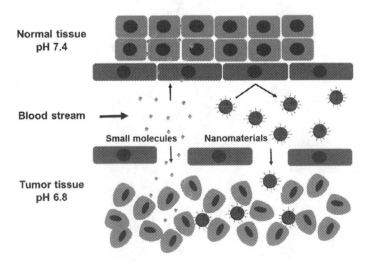

**Fig. 5.1** Schematic illustration of physiological characteristics of normal and tumor tissues with respect to EPR effect for small molecules and nanomaterials

non-specific recognition and uptake by the reticuloendothelial system (RES) is vital to achieve longevity in the blood circulation [15–17]. The size, shape, and surface properties of polymeric nanomaterials are important factors, which determine their biological fate [18–23]. Once reaching the targeted tumor, nanomaterials enable their internalization into the cells via cellular endocytosis. In this stage, a specific ligand was usually introduced into the nanocarriers for elevating the cellular internalization through ligand–receptor interaction with the target cells [24–28]. The active targeting of nanomaterials can realize the further accumulation of anticancer drugs and enhance the therapeutic efficacy.

However, a major obstacle for drug nanocarriers is the inefficient drug release at the tumor site and endosomal entrapment, which significantly reduce anticancer efficacy. To overcome these barriers, external stimuli have been applied to trigger drug release from nanocarriers, such as temperature [29–32], light [33, 34], ultrasound [35], and magnetic field [36, 37]. The effectiveness of these strategies is easily affected by experimental method, tumor location, and size. Exploiting tumor pathophysiology changes including pH [38–42], enzyme activity [43–47], or redox properties [48–51] allows opportunities to develop tumor internal stimuli-responsive nanomaterials. It is well known that there are pH gradients in the tumor sites, and tumor extracellular pH (pH$_e$ ~ 6.5–7.2) is lower than that in blood (pH ~ 7.4). In tumor cells, there are drastically different intracellular microenvironments such as acidic pH in endosomes and lysosomes (pH$_i$ ~ 5–6), reductive microenvironments owing to high level of cysteine or glutathione (GSH) in the cytoplasm, and oxidative microenvironment due to over-generated reactive oxygen species (ROS) in mitochondria (Fig. 5.2). Therefore, acid- and redox-responsive nanomaterials are reported extensively for targeted and controlled anticancer drug delivery.

**Fig. 5.2** Schematic illustration of drug-loaded polymeric nanomaterials and their stimuli-responsive drug release in intracellular microenvironments

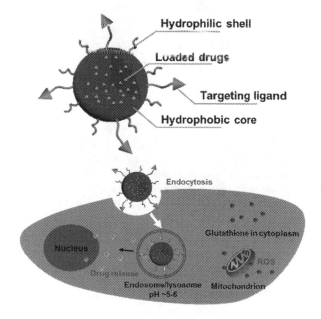

Several nanocarriers including inorganic nanoparticles [52–54], liposomes [55–60], micelles [15, 61–63], polymersomes [64–66], nanogels [6, 67–69], and polymer–drug conjugates [14, 70, 71] have been explored to deliver drugs to the tumor sites. Smart polymeric carrier is one of the simplest and most successful approaches owing to the tunable nanoparticle sizes, diverse chemical modifications, and feasible introduction of stimuli sensitivity [72, 73]. Polymeric nanomaterials also exhibit high stability with a slow rate of dissociation due to quite low critical aggregation concentration (CAC), allowing retention of payloads in nanocarriers for site-specific drug delivery and release. In general, drugs are incorporated into polymeric nanoaggregates by physical entrapment or by chemical conjugation with hydrophilic shell, facilitating prolonged circulation in the bloodstream (Fig. 5.2). Once nanomaterials arrive at tumor sites and enter cells through cellular endocytosis, the physical stimuli (pH, oxidative/reductive potential) would induce the structural changes of polymeric nanocarriers, resulting in rapid release of loaded drugs and exertion of therapeutic effect on tumor cells [73–77].

In this review, the strategies to design acid- and redox-responsive smart polymeric nanomaterials as anticancer drug carriers are discussed. The mechanisms of polymer disassembly and drug release in response to the low pH and reductive/oxidative potential are fully elucidated. Then, the focus was placed on the relationship between chemical structure, stimuli-responsive bonds, self-assembly property, and drug release behavior of polymeric nanomaterials, which may offer a guideline for optimal polymeric nanocarrier design. Finally, we present an outlook and possible challenges in this field of smart polymeric nanomaterials responsive to tumor cellular pH and redox microenvironments.

## 5.2  Acid-Responsive Polymeric Nanomaterials

It is well known that pH difference exists in human bodies widely, and hence, pH values can be applied as a potential trigger to realize site-specific drug delivery. In the blood circulation and normal tissues, the pH was neutral (pH $\sim$ 7.4), while the mild acidic pH was encountered in tumor microenvironment (pH $\sim$ 6.8) as well as in the endosomal and lysosomal compartments of cells (pH $\sim$ 5–6) [78–81]. Therefore, researchers developed acid-sensitive nanocarriers, which can release the systemically administered drugs in specific sites, resulting in targeted and effective drug delivery. In cancer treatment, acid-sensitive polymeric nanomaterials have been extensively studied as drug delivery systems due to their attractive properties such as prolonged blood circulation time and acid-triggered drug release in tumor sites. Generally, the polymers self-assemble into nanoscaled micelles, vesicles, and nanoparticles at pH 7.4, and the anticancer drugs are encapsulated into the nanomaterials to form nanodrugs. In tumor sites or endosome/lysosome compartments, the loaded drugs are released due to the destabilization and dissociation of nanomaterials, realizing the targeted tumor treatment. In the past decades, various

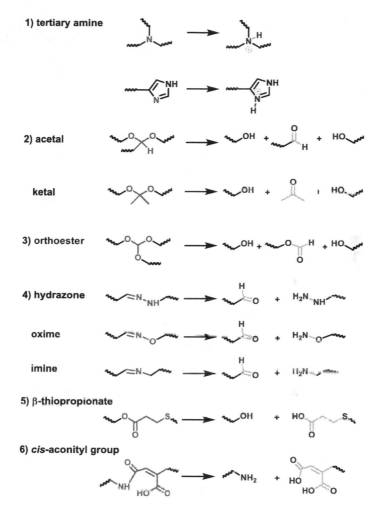

**Fig. 5.3** Acid-responsive motifs and their hydrolyzed products

pH-responsive polymers with different groups/linkers have been reported, such as tertiary amine group, acetal/ketal, orthoester, hydrazone/oxime/imine, β-thiopropionate, and cis-aconityl group (Fig. 5.3).

## 5.2.1  Polymers with Tertiary Amine Group

The "protonation" approach is one of the most important strategies for developing acid-sensitive polymers. The tertiary amine groups were induced into the polymer chains, which are uncharged/hydrophobic at a pH above the $pK_b$ (pH 7.4) and become charged/hydrophilic at a pH below the $pK_b$. The biodegradable polypeptides were selected as pH-sensitive copolymer blocks due to their good biocompatibility. Bae et al. have reported a series of poly(L-histidine) block copolymers with tertiary amines in the pendent groups, which formed micelles by hydrophobic interaction and loaded anticancer drug DOX at neutral pH [82–85]. As the pH became mildly acidic (pH 6.8), the ionization of the tertiary amine causes increased hydrophilicity and electrostatic repulsions of the polymers, leading to the destabilization of micelles and subsequent release of the loaded drugs. The copolymers with desirable pH sensitivity could be applied to controlled drug delivery system that was specifically responsive to the extracellular pH of acidic solid tumors and exhibited effective anticancer activity toward tumor cells. Furthermore, they modified transactivator of transcription (TAT) on the block copolymers, which self-assembled into micelles loading DOX as the model drug [86]. The TAT was shielded during blood circulation and exposed at tumor microenvironment to facilitate the cell internalization process. The DOX-loaded micelle displayed effective tumor suppression activity in vivo.

Methacrylate-based block copolymers with tertiary amine were also reported as pH-sensitive anticancer carriers [87–89]. Shen et al. have synthesized amphiphilic block copolymers by using atom transfer radical polymerization (ATRP), and the anticancer drug cisplatin was loaded into the pH-responsive cores of copolymer nanoparticles [90]. After the nanoparticles were internalized by cells and transferred to lysosomes (pH < 6), the nanoparticles dissolved due to the protonation of amine groups, followed by the rapid release of cisplatin into cytoplasm, resulting in higher cytotoxicity than the pH-non-responsive nanoparticles. Then, the poly($\varepsilon$-caprolactone) (PCL) hydrophobic cores were added to overcome the problems of polymer–drug carriers such as premature burst release of drugs, slow cellular uptake, and undesirable retention in acidic microenvironment of intracellular compartments [91].

Poly($\beta$-amino ester)s (P$b$AE) were originally synthesized by R. Langer et al. via Michael addition of the diamine and diacrylate [92, 93]. A library of P$b$AEs was prepared by combinatorial synthesis, and high-throughput screening approach was explored for discovering efficient polymeric nanomaterials for gene delivery [94–96]. Some of P$b$AEs were also applied as pH-sensitive nanocarriers for controlled drug release, which have tertiary amine groups with a pK$_b$ value around 6.5 [97, 98]. Recently, Kwon et al. reported a block copolymer PEG-$b$-P$b$AE that self-assembled into micelles at neutral pH, loading the anticancer drug DOX (Fig. 5.4a) [99]. At pH 6.4, the loaded DOX was released quickly owing to the increased hydrophilicity of P$b$AE block, which resulted in high antitumor efficiency. The anticancer drug camptothecin (CPT) was also loaded into the micelles, exhibited significantly increased therapeutic efficacy than the control polymeric micelles of PEG-poly(L-lactic acid) (PEG-PLLA) in vivo (Fig. 5.4b) [100]. Chen et al. also synthesized a series of PEG-$b$-P$b$AE copolymers with tunable pK$_b$ values for intracellular drug delivery [101].

In order to enhance the anticancer efficiency of P$b$AE copolymers, the targeted moieties were incorporated. Wang et al. developed a series of poly(RGD-$co$-$\beta$-amino ester) copolymers via one-pot synthesis method which was simple and reliable (Fig. 5.5a) [102]. The copolymers formed nanoparticles at pH 7.4 in aqueous solution with hydrophobic microdomains in the interior and targeting RGD

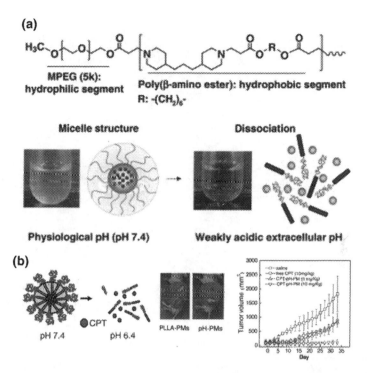

**Fig. 5.4  a** Structure of PEG-b-P*b*AE block copolymer and their self-assembly into micelles at pH 7.4, which were completely dissociated at weakly acidic extracellular pH. Reproduced with permission from Ref. [99]. **b** CPT-encapsulated micelles under neutral pH and rapid release of CPT at pH 6.4 for tumor targeting imaging and therapy in vivo. Reproduced with permission from Ref. [100]

on the surfaces, and DOX was loaded into copolymer nanoparticles by hydrophobic interactions. After the DOX-loaded nanoparticles with RGD were delivered into $\alpha_v\beta_3$ positive U87 cells efficiently, the triggered release of DOX in intracellular acidic compartments results in the obvious cytotoxicity of U87 cells. Furthermore, PEG-grafted P*b*AEs were also synthesized, which could self-assemble into micelle-like nanoparticles at neutral pH, and natural cyclopeptide RA-V and near-infrared (NIR) fluorescent probe squaraine (SQ) were loaded into the polymeric nanoparticles (Fig. 5.5b). The RA-V-/SQ-loaded nanoparticles could deliver RA-V into cancer cells by cellular endocytosis and release the cargos in lysosome efficiently, resulting in the mitochondria disruption and subsequent tumor cell apoptosis in vitro and in vivo [103].

Except for the physical encapsulation of anticancer drugs, the polymer–drug conjugates were also investigated widely. Lee et al. prepared a glycol chitosan grafted with tertiary amine groups, photosensitive drug chlorin e6 (Ce6), and PEG (Fig. 5.6) [104]. At pH 7.4, the polysaccharide–drug conjugates self-assembled into nanoparticles with hydrophobic cores in which Ce6 was encapsulated and

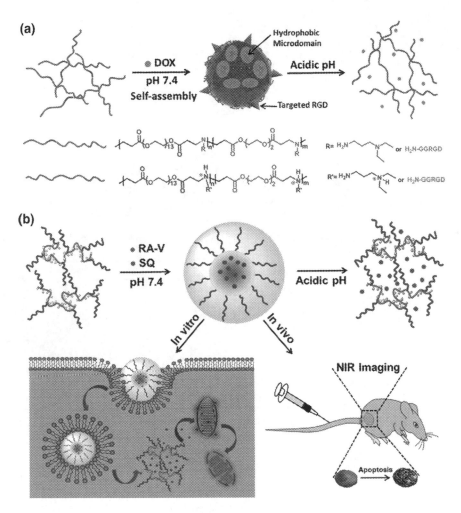

**Fig. 5.5 a** Self-assembly of poly(RGD-*co*-*β*-amino ester) copolymer and its acid-triggered dissociation for efficient drug delivery. Reproduced with permission from Ref. [102]. **b** Co-encapsulation of RA-V and SQ in PEG-grafted P*b*AE micelles for tumor therapy and imaging

autoquenched. When the nanoparticles reached the more acidic surface of the tumor cell (pH 6.8), the protonation of tertiary amine groups induced the dissociation of nanoparticles, and the Ce6 molecules were exposed. Under NIR laser irradiation, the singlet oxygen is generated in tumor site, inducing the cell apoptosis effectively. The conjugate provided targeted high-dose cancer therapy while ensuring the safety of normal tissues. In cancer treatment, peptide drugs have attracted increasing attention. Wang et al. designed cytotoxic peptide (KLAKLAK)$_2$ (named as KLAK)-conjugated PEG-P*b*AEs [105]. The self-assemble polymeric nanoparticles consisted of P*b*AEs as hydrophobic cores and PEG as hydrophilic shells, which

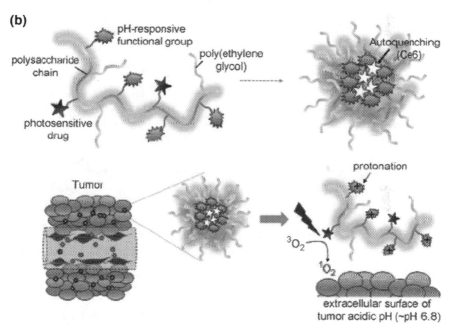

**GCS-g-DEAP-g-Ce6-g-PEG (1)**

**Fig. 5.6  a** Chemical structure of pH-responsive polysaccharide–drug conjugate. **b** Polysaccharide–Ce6 conjugate undergoes autoquenching at neutral pH, and upon reaching the acidic microenvironment of tumor, protonation of polymer occurs and singlet oxygen is generated to kill tumor cells. Reproduced with permission from Ref. [104]

could stabilize the nanoparticles and protect KLAK peptide from enzyme degradation. The chemotherapeutic drug DOX was loaded into the nanoparticle cores, and the drugs were released in mild acidic condition such as endosome or lysosome. The co-delivery of KLAK/DOX and their acid-triggered release resulted in the enhanced anticancer effect in vitro and in vivo. Similarly, the autophagy-inducing peptide Bec1 was conjugated onto the PEG-P$b$AE backbone to gain P-Bec1, and dissociated P-Bec1 can escape from lysosomes and effectively induce autophagy, leading to autophagic cell death [106].

The pH-responsive polymers with tertiary amine groups widely studied in biomedicine fields because of their fast protonation–deprotonation transition and high pH sensitivity. The $pK_b$ values of copolymers can be tuned across a wide range by changing the compositions and chemical structures of the tertiary amine moieties. However, the existence of positive electricity on amine groups may cause the binding of negatively charged protein in blood circulation, leading to undesirable clearance or blotting in blood and high acute toxicity in vivo. Therefore, the proper surface modification was important for enhancing the compatibility of polymeric nanomaterials.

## 5.2.2 Polymers with Acetals/Ketals

Polymers with acid-cleavable bonds were developed extensively as drug carriers, which are stable at neutral pH and hydrolyzed quickly in acidic condition. Frechet et al. synthesized linear-dendritic block copolymers containing PEG hydrophilic chain and pH-sensitive cyclic acetals modified dendrimer block (Fig. 5.7) [107, 108]. The copolymers self-assembled into micelles and loaded anticancer drug DOX by hydrophobic interaction and π–π stacking at neutral pH. As the pH became acidic, the acetals were hydrolyzed and the hydrophilicity of dendrimer block increased, causing the dissociation of micelles and subsequent DOX release. It is worth noting that three methoxy groups in the *ortho* and *para* positions relative to the acetal are important for the acid-responsive hydrolysis of acetal groups due to their electron-donating function [109]. The linear copolymers with acetals/ketals on the main chains were also reported, which formed nanoparticles and loaded with drug and functional proteins, and hence, the nanoparticles could be applied as potential nanocarriers for acid-triggered drug release [110–113].

Recently, block copolymers with acetals/ketals as side groups were also reported for acid-triggered drug release. Zhong et al. prepared pH-responsive amphiphilic copolymers via ring-opening polymerization with acid-labile polycarbonate as hydrophobic block, which self-assembled into micelles at pH 7.4, loading hydrophobic PTX or DOX. As the pH lowered to 5, the fracture of acetals induced the hydrophilicity increase of acid-labile block, resulting in the controlled drug release [114]. Furthermore, the aggregates morphology could be regulated by changing the length of polycarbonate block that could be degraded into small molecules in physiological condition [115]. Acid-responsive cross-linked nanoparticles with acetal side groups were applied as PTX nanocarriers (~ 100 nm). At pH 5.0, the nanoparticles expanded to ~ 1000 nm, simultaneously releasing the loaded drug and suppressing the tumor growth in vivo (Fig. 5.8) [116].

The cationic polymers with ketals as side groups were utilized as tunable non-viral gene carrier for controlled delivery of plasmid DNA and siRNA [117, 118]. Frechet et al. developed acetal-derivatized dextran, forming stable

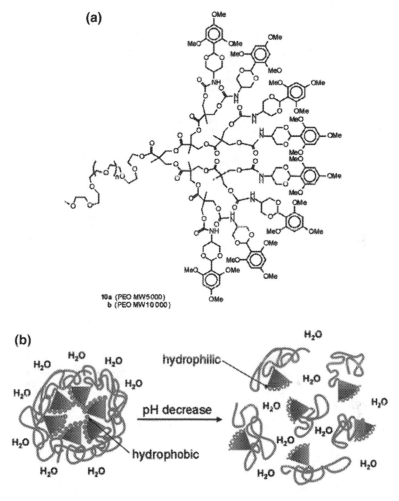

**Fig. 5.7 a** Chemical structure of linear-dendritic block copolymer with cyclic acetals. **b** Schematic for pH-sensitive micelle formation and acid-triggered dissociation due to the hydrolysis of cyclic acetals. Reproduced with permission from Ref. [107]

nanoparticles and encapsulating functional proteins or plasmid DNA at neutral pH [119–121]. The nanoparticles with acid-triggered release properties could be applied for particulate immunotherapy and gene delivery.

## 5.2.3  Polymers with orthoesters

The drug nanocarriers enter cells by endocytosis, forming endosome which then fuses with lysosomes. In this process, the drugs must be released in time to avoid degradation and exocytosis. Considering the limited time in endosome/lysosome,

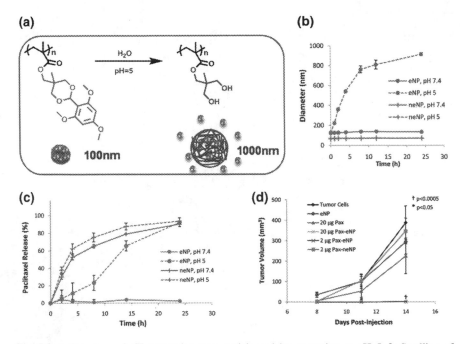

**Fig. 5.8 a** Structure of pH-responsive nanoparticle and its expansion at pH 5. **b** Swelling of nanoparticles as a function of pH and time at 37 °C. **c** Release profiles of paclitaxel from nanoparticles at different pHs. **d** Tumor volume changes over the course of treatments for animals receiving the different treatment groups. Reproduced with permission from Ref. [116]

the sensitivity of nanocarriers should be enough to release drugs in this stage. Compared with acetals/ketals, orthoester groups exhibit faster hydrolysis rate in acidic condition, and hence, polymers with orthoesters may be more practical for intracellular drug release.

Poly(orthoesters) were first developed as hydrogels for ocular delivery, and then, they were prepared into nanoparticles for gene delivery. The effective protection plasmid DNA in physiological condition and fast release at low pH made the polymers into potential gene carriers [122, 123]. Recently, a family of acid-labile polymers with orthoesters side groups were studied by Li et al. for controlled drug delivery [124, 125]. The lower critical solution temperatures of thermoresponsive polymers increased with the hydrolysis of the pendent orthoester groups [126]. And then, they prepared a series of block copolymers composed of a PEG block and a thermoresponsive and acid-labile block [127]. The copolymers could form micelle-like nanoparticles by directly heating the copolymer solution above the critical aggregation temperature (CAT). By increasing the relative chain length of PtNEA to PEG, the copolymers could spontaneously form polymersomes, encapsulating both hydrophobic drugs and water-soluble biomacromolecules (Fig. 5.9) [128]. The acid-triggered release of payloads made the copolymers as promising carriers of multi-drugs for site-specific delivery. Wang et al. also designed block

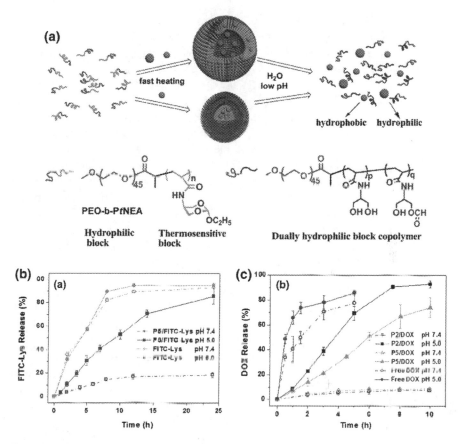

**Fig. 5.9** **a** Thermally induced nanoparticle and polymersome formation and their acid-triggered dissociation. **b** pH-dependent cumulative release of FITC-Lys from polymersomes. **c** pH-dependent cumulative release of DOX from nanoparticle

copolymers bearing acid-labile orthoester side chains, and the DOX-loaded micelles exhibited the enhanced drug delivery to human glioma cells [129].

## 5.2.4   Polymers with Hydrazone/Oxime/Imine

Aldehydes or ketones can condense with molecules containing amine groups, such as hydrazine, hydroxylamines, and primary amines, resulting in a class of compounds known as hydrazones, oximes, and imines, respectively. These linkages are stable at physiological pH due to the participation of $\pi$ electrons in an electron delocalization effect and resultant stable carbocation. The fracture of these chemical bonds in mild acidic condition endows the compounds acid-labile property. Hydrazone bond is commonly used to conjugate anticancer drugs with polymer main chains. Ulbrich et al. first developed various acid-labile polymer–drug conjugates with hydrazone bonds as sensitive spacers [130, 131]. The anticancer drug DOX was covalently attached via hydrazone bond to water-soluble N-(2-hydroxypropyl)methacrylamide (HPMA) copolymer–drug carriers, which showed relative stability at pH 7.4 and a fast DOX release at pH 5. The positively or negatively charged groups or hydrophobic substituents had an effect on the DOX release behaviors. Furthermore, they prepared star-shaped immunoglobulin-containing polymer–DOX conjugates and conjugates of DOX with graft HPMA copolymers, which exhibit prolonged blood circulation and enhanced tumor accumulation and suppression [132, 133]. Both DOX and PTX were also conjugated onto the HPMA main chains, and the drug release profiles were related to the linkage groups and amount of covalent drug [134]. The star HPMA copolymer nanocarriers were also designed for simultaneous anticancer drug delivery and imaging [135]. Recently, HPMA copolymer-conjugated pirarubicin has been applied in multimodal treatment of a patient with stage IV prostate cancer and extensive lung and bone metastases [136]. The anticancer drug DOX was also conjugated onto polypeptide–PEG block copolymers via hydrazone bonds to gain PEG-p(Asp-Hyd-ADR), which self-assembled into nanoscaled micelles by hydrophobic interaction (Fig. 5.10a). The micelles remained stable at pH 7.4–7.0 and released the loaded drugs under acidic conditions below pH 6.0 (Fig. 5.10b). Therefore, the micelles could realize intracellular pH-triggered drug release, exhibiting tumor-infiltrating permeability and effective antitumor activity with extremely low toxicity in vitro and in vivo [137, 138]. The amphiphilic multi-arm block copolymers conjugated with DOX via hydrazone bonds were also developed for tumor-targeted drug delivery [139]. In order to realize rapid cellular uptake and efficient drug release in tumor cells, poly(choline phosphate) prodrugs were designed for rapid internalization by cancer cells due to the strong interaction between multivalent choline phosphate groups and cell membranes [140]. The disulfide bonds were introduced into the cores of micelles, producing DOX-conjugated acid and reduction dual-sensitive micelles [141, 142]. For the cross-linked prodrug nanogels, highly selective intracellular drug delivery and upregulated antitumor efficacy were investigated in vitro and in vivo [143]. In addition, antibody–polymer conjugates with oxime as linkages were studied for selective delivery of small siRNA to target cells [144].

For pH-controlled drug delivery, the anticancer drugs also could be linked onto the end of polymer chains. For example, DOX was attached onto the hydrophobic end of PLA–PEG block copolymer and simultaneously loaded into their formed micelles [145]. At acidic media, the hydrolysis of hydrazone bond induced the pH-dependent release of DOX. Yang et al. synthesized PEG-DOX conjugates with benzoic-imine linker, which could co-assemble with other amphiphilic graft copolymers, forming pH-sensitive polymeric vesicles for acid-triggered DOX release [146]. Furthermore, the micelles were prepared by combining PEG-DOX with targeting RGD-modified block copolymer, and mild acidic conditions caused the cleavage of PEG-DOX, exposing the RGD to guide the micelle to target cancer cells [147].

**Fig. 5.10 a** Preparation of drug-loaded polymeric micelles from self-assembling amphiphilic block copolymers PEG-p(Asp-Hyd-ADR), in which DOX was conjugated through hydrazone linkers. **b** DOX release profiles at different pHs. Reproduced with permission from Ref. [138]

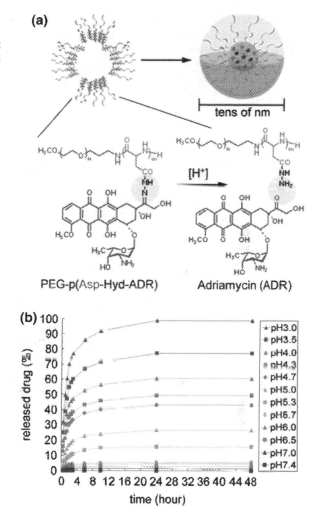

Besides the polymer–drug conjugates, the hydrazone/oxime/imine bonds were widely used as polymer linkers. Torchilin et al. first reported amphiphilic PEG–phosphatidylethanolamine (PE) with hydrazone as acid-labile linker between PEG block and PE block [148]. The surfaces of PEG–PE nanocarriers were further modified by biotin or TAT peptide moieties, which were shielded by long protecting PEG chains in physiological condition. At lower pH (5–6), nanocarriers lost their protective PEG shell because of acidic hydrolysis of hydrazones, and the functional biotin or TAT was exposed for effective internalization into cells. Similarly, PEG chains were grafted onto the polycationic main chains via benzoic-imine linkers, producing pH-triggered reversible stealth micelles. Imine was also utilized as a linker between hydrophobic block and hydrophilic block [149]. The block copolymers formed micelles loading DOX at physiological pH, and the acidic condition resulted in the cleavage of imine bond and the ionization of micelle surface, facilitating the cellular uptake of the micelles through the electrostatic interaction. In addition, a series of copolymers with oxime bonds were first synthesized by Wooley et al. [150]. Then, Miller et al. developed oximes containing gene nanocarriers, which could combine siRNA for controlling HBV replication in vivo [151]. The pH-triggered nanoparticles could also realize the siRNA delivery to liver cells in vitro and in vivo [152].

### 5.2.5  Polymers with β-Thiopropionate Group

The reaction of thiol group with acrylate group yields the β-thiopropionate linkage, which has been discovered with acid-labile property. The group can be hydrolyzed selectively under mild acidic conditions at a much slower rate than other acid-labile functional groups. Kataoka et al. first prepared an oligodeoxynucleotide(ODN)–PEG conjugate through β-thiopropionate linkage [153]. The block copolymer was self-assembled with linear PEI, forming stable polyion complex micelle at pH 7.4. The ODN–PEG conjugate showed cleavage of the ester linkage at the endosomal pH (5.5), suggesting that ODN could be released from the micelle in the intracellular compartment. Then, they further synthesized lactosylated PEG–siRNA conjugate with β-thiopropionate linkage, constructing pH-sensitive polyion complex micelles achieving enhanced gene silencing in hepatoma cells [154]. In addition, amphiphilic triblock copolymers were reported with hydrophobic block containing β-thiopropionate groups, which self-assembled into vesicles at pH 7.4 and released the loaded guest molecules at pH 5.3 (Fig. 5.11) [155].

### 5.2.6  Polymers with Cis-Aconityl Group

Polymers with amine groups can react with carboxy dimethylmaleic anhydride, yielding maleic anhydride derivatives with acid-labile cis-aconityl group, which was widely used as linkage between nanocarriers and protective PEG shell. For example, siRNA dynamic polyconjugate including cis-aconityl-PEG was reported for effective cellular uptake and disassembly in the low pH environment of the endosome [156]. The release of the siRNA into the cytoplasm of the target cell enabled in vivo delivery of siRNA to hepatocytes. Kataoka et al. exploited the cis-aconityl group to develop charge-conversional polyionic complex micelles as efficient nanocarriers for gene and protein delivery into cytoplasm [157, 158]. The siRNA was also conjugated onto the copolymer main chains via cis-aconityl group, and the cleavage of siRNA from polymer chains in endosome accelerated the endosomal escape and siRNA release [159]. For anticancer drug delivery, PEGylated glycolipid conjugate DOX-loaded micelles were reported, and the cleavage of PEG shell at tumor microenvironment exhibited excellent cellular uptake ability into cells [160]. Except for the physical encapsulation, the DOX

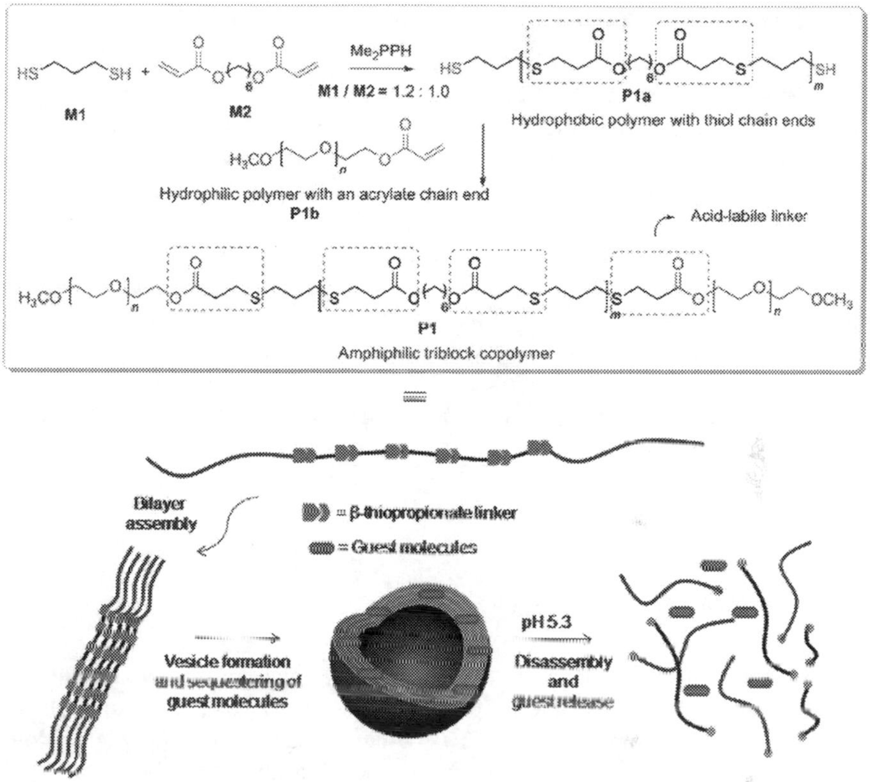

**Fig. 5.11** Synthesis route of amphiphilic triblock copolymer with β-thiopropionate groups and its pH-responsive disassembly. Reproduced with permission from Ref. [155]

could also be linked onto linear polymer or dendrimer through cis-aconityl linkage [161–164]. The polymer–DOX conjugates were stable at blood circulation and release the conjugated DOX at tumor site, realizing the controlled and targeted drug delivery in vivo. Wagner et al. created a protein transduction shuttle with cis-aconityl linkage for intracellular delivery and release of active proteins [165]. Recently, Wang et al. reported polymeric clustered nanoparticle (iCluster) with an initial size of ~100 nm, and the platinum drug-conjugated dendrimer PAMAM/Pt was linked onto the surface of iCluster through cis-aconityl linkage [166]. Once iCluster accumulates at tumor sites, the tumor extracellular acidity triggered the release of PAMAM/Pt (diameter ~5 nm), which facilitated tumor penetration and cell internalization of the anticancer drugs.

## 5.3 Reduction-Sensitive Polymeric Nanomaterials

It has been demonstrated that glutathione (GSH) is the most abundant biological thiol in animal cells. GSH/glutathione disulfide (GSSG), as the major redox couple, regulates the microenvironment potentials by their ratio variety. In blood and extracellular matrices, there is a relatively high redox potential due to a low concentration of GSH (approximately 2–20 μM). However, the cytosol and nuclei maintain a highly reducing environment with the GSH concentration of 0.5–10 mM under the reduction of NADPH and glutathione reductase. Meanwhile, researchers have demonstrated that the tumor tissues showed at least fourfold higher GSH concentrations than that of normal tissue in mice, rendering the reduction-sensitive nanomaterials valuable for tumor-specific drug delivery. The polymers and conjugates with disulfide linkages in the main chain or cross-linker usually exhibit reduction-sensitive property.

### 5.3.1 Reduction-Sensitive Polymer with Disulfide in Main Chain

In block copolymer, the disulfide bonds are usually designed to link hydrophilic and hydrophobic block. For example, Zhong et al. developed disulfide-linked dextran-b-poly(ε-caprolactone) (PCL) diblock copolymer, which could self-assemble into reduction-responsive biodegradable micelles and load hydrophobic DOX [167]. After endocytosis by cells, GSH triggered dextran shells shed off through the cleavage of the intermediate disulfide bond in the presence of glutathione tripeptide, which results in fast destabilization of micelles and subsequent release of DOX in the cytoplasm as well as to the cell nucleus. The ligand galactose was modified onto the surface of PCL–PEG micelles for effective targeting to hepatocellular carcinoma cells and controlled drug release in reductive intracellular microenvironment [168]. The targeting peptide cRGD was also linked on the micelle to realize enhanced doxorubicin delivery to human glioma xenografts in vivo [169]. The cystamine diisocyanate could be used as a coupling agent for synthesis of reductively degradable biopolymers with disulfide bonds on the main chain [170]. Furthermore, acetal groups were incorporated into the disulfide-linked diblock copolymer, obtaining reduction and pH-responsive degradable micelles for dually activated intracellular anticancer drug release [171]. Thayumanavan et al. reported triple sensitive copolymers containing thermoresponsive block PNIPAM and acid-labile block with acetal group linked by disulfide, which not only showed synergistic effects, but also provided multiple mode control of the resulting micelle system [172].

## 5.3.2   Reduction-Sensitive Polymer with Disulfide as Cross-linker

For polymer with disulfide as cross-linker, one of the most common strategies was the introduction of disulfide-containing cross-linkers. Li et al. prepared multi-responsive nanogels cross-linked by disulfide with OEG side chains and pendent orthoester groups through miniemulsion polymerization [173]. These nanogels are thermoresponsive and labile in the weakly acidic or reductive environments. The hydrophobic drugs DOX and PTX were loaded into the nanogels, and the drug release from the nanogels can be greatly accelerated by a cooperative effect of both acid-triggered hydrolysis and DTT-induced degradation. Matyjaszewski prepared biodegradable nanogels cross-linked with disulfide linkages by inverse miniemulsion ATRP (Fig. 5.12) [174, 175]. The nanogels with a uniformly cross-linked network could encapsulate molecules including rhodamine 6G, a fluorescent dye, and DOX, and the nanogels biodegraded into water-soluble polymers in the presence of glutathione, resulting in the controlled release of loaded cargoes.

Except for the introduction of disulfide-containing cross-linkers, Thayumanavan et al. synthesized self-cross-linked nanogels by a simple thiol–disulfide exchange reaction (Fig. 5.13) [176]. Nanogels with controlled size and guest release rate formed after adding a certain amount of DTT into the precursor solution. The payload encapsulation stability of nanogels and reduction-triggered release has been demonstrated through fluorescence resonance energy transfer (FRET) experiments (Fig. 5.13c). The non-covalently encapsulated guest molecules such as dyes and DOX can be released in response to GSH. Shuai et al. developed disulfide cross-linked micelles through the oxidation of thiols in polymer side chains [177]. The drug-loaded micelles were stable in blood circulation, and a burst release of drug was triggered in an acidic and reductant-enriched environment. Cell and animal studies revealed the great potential of the dual-sensitive micelles in cancer treatment. Zhong et al. exploit cystamine as the cross-linker to obtain in situ

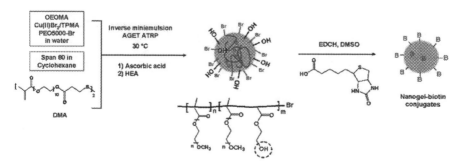

**Fig. 5.12** Synthesis and modification of biodegradable nanogels cross-linked with disulfide linkages. Reproduced with permission from Ref. [174]

forming reduction-sensitive degradable nanogels, realizing the facile loading and triggered intracellular release of proteins [178]. They also reported core-cross-linked polypeptide micelles through similar disulfide exchange reaction, which cause intracellular release of doxorubicin in both pH and reduction conditions [179, 180].

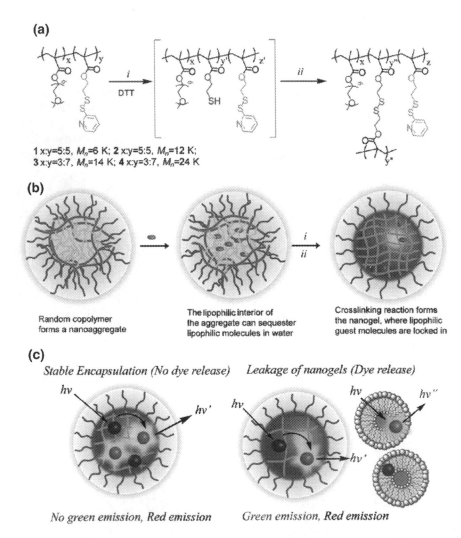

**Fig. 5.13 a** Chemical structure of the polymer precursors and nanogels with disulfide cross-linking. **b** Schematic representation of the preparation of the cross-linked nanogels. **c** FRET experiment for the stable and leaky nanogels. Reproduced with permission from Ref. [176]

## 5.4   Oxidation-Responsive Polymeric Nanomaterials

ROS and reactive nitrogen species (RNS), which are two types of oxidative species, widely exist in the human body and play crucial roles in many physiological and pathological processes such as cell signaling, proliferation, apoptosis, and host innate immunity [50, 181]. ROS mainly include hydrogen peroxide ($H_2O_2$), superoxide anion ($O_2 \cdot^-$), hydroxyl radical (•OH), hypochlorite ($OCl^-$), peroxyl radicals, and singlet oxygen, which are generated from several endogenous sources, mainly in the mitochondria, from an incomplete reduction of NADPH and oxygen in the plasma membrane [182, 183]. RNS generally include nitric oxide and its derivatives, such as nitrogen dioxide and peroxynitrite ($ONOO^-$) [184]. ROS or RNS at appropriate concentrations are strongly related to metabolic activities and are involved with a number of cell functions related to both health and disease. During the normal metabolic activity, ROS are continuously produced, transformed, and consumed in cells or tissues to maintain a complex balance between ROS generation and elimination. An appropriate concentration of ROS is involved with normal biological functions in many physiological processes [185]. However, excessive amounts of ROS or RNS result in oxidative stress and impairing the critical biomolecules of cells including lipids, proteins, and DNA [186]. Oxidative stress is strongly related to a series of pathophysiological events, such as cancers, cardiovascular, inflammatory, atherosclerosis, diabetes, and neurodegenerative diseases [187, 188]. Therefore, ROS-sensitive nanomaterials have attracted considerable attention for promising drug delivery system by targeting the specific oxidative microenvironments of disease sites. In the past decade, a series of oxidation-responsive polymers have been reported as potential candidates for site-targeted drug delivery.

We present a summary of recent progress of the various oxidation-responsive polymers that can be applied as potential drug nanocarriers. Oxidation-responsive polymers possess unique properties including the transformation of solubility or the degradation of polymers caused by the oxidation of ROS. The chemical structures of representative types of ROS-sensitive polymers are shown in Fig. 5.14, including poly(propylene sulfide) (PPS), selenium-containing polymers, phenylboronic ester-containing polymers, aryl oxalate-containing polymers, poly(thioketal)s, and proline-containing polymers.

### 5.4.1   Poly(Propylene Sulfide)S

The sulfur(II)-based oxidation-responsive polymers (i.e., poly(alkylene sulfide)s and poly(arylene sulfide)s) are one type of the most explored ROS-responsive materials. Poly(alkylene sulfide)s based on the triblock copolymer of poly(propylene sulfide) (PPS) and PEG were originally reported as drug vehicles by Hubbell et al. in 2004 [189]. The PPS possesses much more hydrophobicity than its oxygen

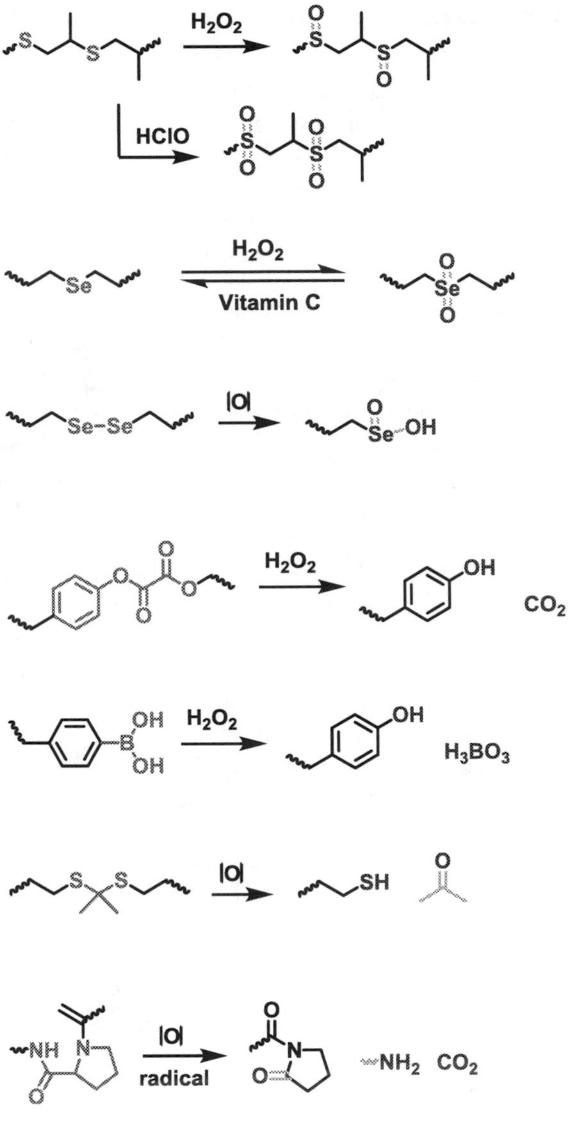

**Fig. 5.14** Oxidation-responsive motifs and their oxidative products

analogue (i.e., PEG) owing to its low dipole moment, which enables the copolymers to self-assemble into unilamellar vesicles. The vesicles dissociated due to the variation of solubility from hydrophobicity to hydrophilicity under oxidative conditions, which was caused by the generation of polysulfoxide and/or polysulfone. These unique properties endow the PPS-based copolymers as drug/gene delivery system specifically responsive to the oxidation-related disease sites.

In recent studies, other groups systematically explored a series of PPS-based polymers as potential candidates of drug delivery specifically responsive to the oxidation microenvironment in the forms of vesicles, micelles, and cross-linked nanoparticles. Hydrogen peroxide ($H_2O_2$)-responsive hyperbranched polymers used as effective anticancer micelle system were reported by Zhu et al. [190]. The hydrophobic anticancer drug 7-ethyl-10-hydroxy-camptothecin (SN38) was linked to hydrophilic hyperbranched polyglycerol (HPG) via thioether linkage to form amphiphilic hyperbranched polymers HPG-2S-SN38 (Fig. 5.15). The copolymers could self-assemble into stable micelles, and anticancer drug cinnamaldehyde (CA) was effectively loaded into them. The micelle was easily disrupted due to the oxidation of thioether, leading to the release of SN38 and CA. The synergistic anticancer effect of SN38 and CA resulted in amplified antitumor effect via enhancing the cell apoptosis, suggesting that these nanomaterials are promising candidates as novel anticancer therapeutics. Duvall et al. also reported a PPS-based micelle for ROS-triggered hydrophobic drugs/dyes release in the site of inflammation and oxidative environments (i.e., $H_2O_2$, 3-morpholinosydnonimine (SIN-1), and peroxynitrite) [191]. The PPS-based polymer self-assembled into micelles with properties of oxidation-responsive disassembly and cargo release, indicating that they have great potential for drug release under physiological and pathological

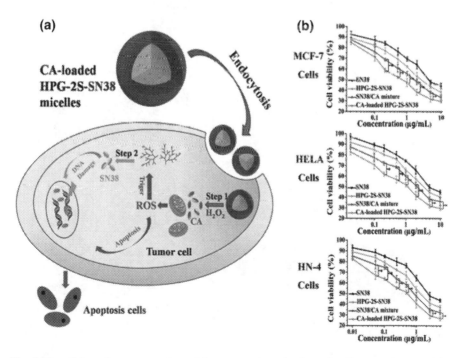

**Fig. 5.15** **a** Schematic representation of the proposed mechanism of cell apoptosis caused by CA-loaded PPS-based polymer micelles. **b** Cell viability of CA-loaded micelles in MCF-7, Hela, and HN-4 cells. Reproduced with permission from Ref. [190]

conditions. Recently, Duvall et al. fabricated PPS into microparticles and successfully utilized it for encapsulation and "on-demand" delivery of curcumin in a model of diabetic peripheral arterial disease [192]. Owing to the ROS scavenging ability of PPS, the blank microparticles also exhibited therapeutic properties. The combination of PPS with curcumin established the utility of this platform for reducing cell death under oxidative stress of chronic inflammatory diseases.

Poly(β-thioester)s was also developed as ROS-responsive polymeric nanocarriers. Wang et al. conjugated KLAK onto amphiphilic poly(β-thioester)s copolymers (H-P-K), which self-assembled into micelles loading SQ dye (Fig. 5.16) [193]. The H-P-K nanodrug exhibited high antitumor activity due to the efficient disruption of mitochondrial membranes. The excessive ROS production in apoptosis process stimulated the swelling of nanoparticles and subsequent release of SQ, resulting in the significant increase of photoacoustic (PA) signal. The polymeric nanomaterial has great potential for high antitumor performance and in situ cancer treatment evaluation. Additionally, poly(β-thioester)s has also been developed to combine with other stimuli-sensitive (thermal, pH, GSH) moieties to obtain multi-responsive systems. In a recent example of oxidation and thermal dual-responsive systems reported by Chen et al., the dual-responsive behavior makes these copolymers promising for the release of the cargoes in specific diseased regions [194]. ROS and pH dual-responsive poly(β-thioester)s with grafted PEG chains were also prepared, realizing the controlled DOX loading and intracellular release [195].

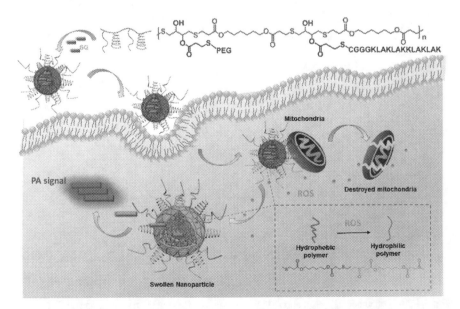

**Fig. 5.16** ROS-sensitive polymer–KLAK conjugates self-assembled into micelle-like nanoparticles for enhanced tumor treatment and in situ therapeutic performance evaluation

## 5.4.2 Selenium-Based Polymers

Similar to the sulfide groups in thioether-based polymers, the selenium also possesses oxidation-responsive properties. Under oxidative environment, hydrophobic selenides are able to be easily oxidized to water-soluble selenoxides and selenones [50]. However, selenium(II)-based polymers possess a higher sensitivity to oxidants than sulfur(II)-based analogues due to the weaker bond energy of the C–Se bond compared with C–S bond [196]. Due to these features, selenium-containing polymers are potential candidates as oxidation-responsive drug carriers or antioxidant materials. Zhang et al. have reported a series of ROS-responsive selenium-based nanoparticles for drug delivery [197]. These nanoparticles formed from the main-chain-type or monoselenium-containing copolymer underwent a structural disassembly and exhibited a triggered release of encapsulated cargos in 0.1 wt% $H_2O_2$ due to the generation of hydrophilic selenoxide. Later, Xu et al. complexed selenide-based copolymer with platinum cation ($Pt^{2+}$) to self-assemble into novel coordination-responsive micelles with DOX loading [198]. These DOX encapsulated in micelles can be released in a controlled manner through the competitive coordination of $Pt^{2+}$ with GSH in a GSH-rich environment. In another strategy, Yan et al. have developed a series of ROS-responsive nanocarrier formed from selenide-based hyperbranched polymer consisting of alternative hydrophobic selenide units and hydrophilic phosphate segments in the polymer backbone [199]. These hyperbranched polymers possess intrinsic antitumor effect by inducing apoptosis of cancer cells. These nanoparticles are capable of loading anticancer drug and releasing it under oxidative or reductive environment, exhibiting synergistic therapeutic effect of cancer.

Additionally, diselenide has great potential for a dual redox response owing to its good activity in the presence of either reductants or oxidants. Se–Se bonds can be reduced to selenol in the presence of reductants or cleaved and oxidized to seleninic acid in an oxidative environment [185]. Zhang et al. reported a dual redox-sensitive micelle formed from Se–Se bonds-based block copolymer (PEG-PUSeSe-PEG), which showed triggered dissociation in an oxidative environment or a reducing environment (Fig. 5.17) [200].

## 5.4.3 Phenylboronic Ester

Phenylboronic acid and phenylboronic ester are important building blocks for fabricating organic compounds and functional polymers [201]. The phenylboronic acid (ester) bonds are susceptibility toward $H_2O_2$, generating phenol and boronic acid as the oxidation products [50]. Frechet et al. prepared oxidation-responsive dextran (Oxi-DEX) carrier microparticles by modifying the hydroxyls of dextran with arylboronic esters [202]. The nanoparticles exhibited a markedly accelerated degradation rate in the presence of $1 \times 10^{-3}$ M $H_2O_2$. Ovalbumin (OVA) as a

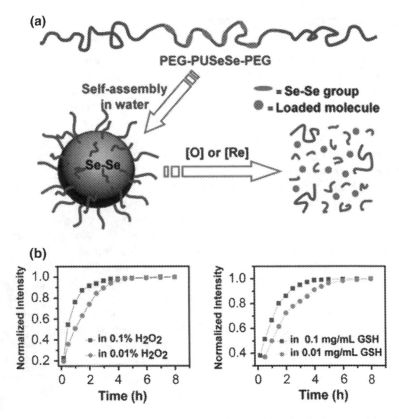

**Fig. 5.17** **a** Schematic of dual redox-responsive disassembly of selenium-containing block copolymer micelles. **b** Payload release profiles at oxidative or reductive condition. Reproduced with permission from Ref. [200]

model vaccine was encapsulated into the microparticles and displayed a 27-fold increase in the presentation to CD8 + T cells relative to OVA loaded in a non-oxidation degradable vehicle. The accelerated degradation of the vaccine molecules release was assumed to be caused by the high levels of $H_2O_2$ in dendritic cells (DCs). Xu et al. have developed a convenient chemical approach based on the phenylboronic acid containing polymers to reversibly modulate protein (RNase A) function, which is sensitive to ROS for targeted cancer therapy [203]. Recently, Liu et al. have reported the oxidation-sensitive polymersomes, which assembled from $H_2O_2$-reactive phenylboronate-based polymers [204]. These polymersomes possess intracellular $H_2O_2$-triggered cross-linking, bilayer permeability switching, and enhanced imaging/drug release features. Furthermore, Li et al. synthesized oxidation and temperature dual-responsive polymers based on phenylboronic acid and N-isopropylacrylamide [205]. The block copolymers formed stable micelle-like nanoparticles and encapsulated DOX by heating to 37 °C, which showed $H_2O_2$ triggered release and obvious cytotoxicity toward cancer cells.

### 5.4.4 Thioketals

Thioketals are able to be easily oxidized to ketones and organic thiols (or disulfide) in the presence of superoxide or some other oxidants [206]. Xia et al. have reported a thioketal-based poly(amino thioketal) (PATK), which is degradable by ROS such as superoxide and hydroxyl radical [207]. The synthesized cationic PATK polymer is capable of condensing DNA to form DNA/PATK polyplexes, exhibiting efficient gene delivery in prostate cancer cells. Under ROS conditions, the degradation of thioketal bonds of PATK led to targeted intracellular release of the combined DNA in cancer cells, resulting in a higher transfection efficiency in cancer cells.

Polymeric nanomaterials that combine diagnostic and therapeutic functions are highly desirable for image-guided cancer therapy. Liu et al. synthesized conjugated polyelectrolytes (CPEs) backbone with PEG side chain modified by DOX through thioketal groups (PFVBT-g-PEG-DOX, Fig. 5.18) [208]. The polyprodrug self-assembled into nanoparticles, and the surface was further functionalized with a cyclic RGD peptide. Under irradiation with light, the polymer generated ROS efficiently for photodynamic therapy. Meanwhile, ROS can quickly cleave the thioketal linker and release on-demand DOX in cells, realizing the combined chemo-photodynamic therapy.

### 5.4.5 Others

Owing to the crucial function in peroxide-based chemiluminescence systems, aryl oxalates have been widely explored since the 1960s [209–214]. In addition to the applications for in vivo $H_2O_2$ imaging, polyoxalate copolymers also have been extensively explored as polymeric antioxidants against oxidative stress-related diseases. Bae et al. have reported $H_2O_2$-responsive antioxidant nanoparticles formed from copolyoxalate containing vanillyl alcohol (VA) (PVAX) as a targeting therapeutic agent for ischemia/reperfusion [215]. PVAX was synthesized by covalently incorporating $H_2O_2$-responsive peroxalate ester linkages and VA in its backbone. PVAX nanoparticles were degraded and VA was released, resulting in anti-inflammatory and anti-apoptotic activity by reducing the generation of ROS. PVAX nanoparticles specifically reacted with over-expressed $H_2O_2$, exhibiting highly potent anti-apoptotic activities and anti-inflammatory ability.

Among stimuli-responsive polymers, polypeptides have played an important role in the biomedical field, owing to their good biocompatibility and biodegradability, as well as the unique properties of mimicking naturally derived proteins. Particularly, polypeptides containing proline residues are subject to degradation by oxidants. In recent study, Sung et al. synthesized a series of ROS cleavable scaffolds containing poly(L-proline), which were cleaved in 6 d in the presence of $5 \times 10^{-3}$ M $H_2O_2$ and $50 \times 10^{-6}$ M Cu(II) [216]. And the scaffolds displayed accelerated degradation by treating with the ROS generator SIN-1. Increased

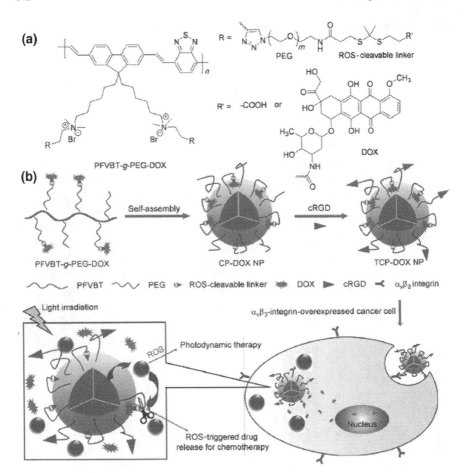

Fig. 5.18 **a** Chemical structure of polyprodrug PFVBT-g-PEG-DOX. **b** Schematic illustration of the light-regulated ROS-activated on-demand drug release for targeted and image-guided photodynamic and chemotherapy. Reproduced with permission from Ref. [208]

scaffold porosity was investigated by co-culturing with activated macrophages, exhibiting a cell-interacting degradation manner. Later, a similar proline-based polymer matrix (KP$_7$K) was developed and an in vivo model applied in long-term tissue engineering was investigated. The submicrometer diameter pores of the KP$_7$K cross-linked scaffold increased by more than tenfold by co-culturing with LPS-activated mouse bone marrow-derived macrophages for 7 d, suggesting a ROS-mediated degradation process.

## 5.5  Summary and Perspective

Considerable efforts concerning smart polymer assemblies have been made over the past decade due to their intrinsic advantages for the cancer treatment. In this review, the chemical structure and drug loading/release property of pH and redox-responsive copolymers are discussed in detail, exhibiting their potential application in controlled drug delivery. Even though great advances have been achieved in the field stimuli-responsive polymers, there still remain several challenges in this field: (1) The sensitivity of polymer aggregates responding to mild acidic condition or low concentrations of the biologically relevant ROS/GSH should be further improved, especially in early cancer therapy and (2) further exploration of the detailed degradation mechanisms of the smart polymers and the fate of their degradation products within cells or in vivo are of great importance in the evaluation of in vivo biocompatibility.

## References

1. Cao Y, DePinho RA, Ernst M, Vousden K (2011) Cancer research: past, present and future. Nat Rev Cancer 11(10):749–754
2. Rothenberg ML, Carbone DR, Johnson DH (2003) Improving the evaluation of new cancer treatments: challenges and opportunities. Nat Rev Cancer 3(4):303–309
3. Langer R (1998) Drug delivery and targeting. Nature 392(6679):5–10
4. Davis ME, Chen Z, Shin DM (2008) Nanoparticle therapeutics: an emerging treatment modality for cancer. Nat Rev Drug Disc 7(9):771–782
5. Brannon-Peppas L, Blanchette JO (2004) Nanoparticle and targeted systems for cancer therapy. Adv Drug Deliver Rev 56(11):1649–1659
6. Chacko RT, Ventura J, Zhuang J, Thayumanavan S (2012) Polymer nanogels: a versatile nanoscopic drug delivery platform. Adv Drug Deliver Rev 64(9):836–851
7. Park K, Lee S, Kang E, Kim K, Choi K, Kwon IC (2009) New generation of multifunctional nanoparticles for cancer imaging and therapy. Adv Funct Mater 19(10):1553–1566
8. Biju V (2014) Chemical modifications and bioconjugate reactions of nanomaterials for sensing, imaging, drug delivery and therapy. Chem Soc Rev 43(3):744–764
9. Phillips MA, Gran ML, Peppas NA (2010) Targeted nanodelivery of drugs and diagnostics. Nano Today 5(2):143–159
10. Ulbrich K, Holá K, Šubr V, Bakandritsos A, Tuček J, Zbořil R (2016) Targeted drug delivery with polymers and magnetic nanoparticles: covalent and noncovalent approaches, release control, and clinical studies. Chem Rev 116(9):5338–5431
11. Schroeder A, Heller DA, Winslow MM, Dahlman JE, Pratt GW, Langer R, Jacks T, Anderson DG (2012) Treating metastatic cancer with nanotechnology. Nat Rev Cancer 12(1):39–50
12. Maeda H, Matsumura Y (1989) Tumoritropic and lymphotropic principles of macromolecular drugs. Crit Rev Ther Drug Carrier Syst 6(3):193–210
13. Maeda H, Wu J, Sawa T, Matsumura Y, Hori K (2000) Tumor vascular permeability and the EPR effect in macromolecular therapeutics: a review. J Control Release 65(1–2):271–284
14. Haag R, Kratz F (2006) Polymer therapeutics: concepts and applications. Angew Chem Int Ed 45(8):1198–1215

15. Kataoka K, Harada A, Nagasaki Y (2001) Block copolymer micelles for drug delivery: design, characterization and biological significance. Adv Drug Deliver Rev 47(1):113–131
16. Nishiyama N, Kataoka K (2006) Nanostructured devices based on block copolymer assemblies for drug delivery: designing structures for enhanced drug function. Adv Polym Sci 193:67–101
17. Duan X, Li Y (2013) Physicochemical characteristics of nanoparticles affect circulation, biodistribution, cellular internalization, and trafficking. Small 9(9–10):1521–1532
18. Champion JA, Katare YK, Mitragotri S (2007) Particle shape: a new design parameter for micro- and nanoscale drug delivery carriers. J Control Release 121(1–2):3–9
19. Albanese A, Tang PS, Chan WCW (2012) The Effect of nanoparticle size, shape, and surface chemistry on biological systems. In: Yarmush ML (ed) Annual review of biomedical engineering, vol 14, pp 1–16. https://doi.org/10.1146/annurev-bioeng-071811-150124
20. Doane TL, Chuang C-H, Hill RJ, Burda C (2012) Nanoparticle zeta-potentials. Acc Chem Res 45(3):317–326
21. Kim ST, Saha K, Kim C, Rotello VM (2013) The role of surface functionality in determining nanoparticle cytotoxicity. Acc Chem Res 46(3):681–691
22. Jo DH, Kim JH, Lee TG, Kim JH (2015) Size, surface charge, and shape determine therapeutic effects of nanoparticles on brain and retinal diseases. Nanomed Nanotechnol Biol Med 11(7):1603–1611
23. Salatin S, Maleki Dizaj S, Yari Khosroushahi A (2015) Effect of the surface modification, size, and shape on cellular uptake of nanoparticles. Cell Biol Int 39(8):881–890
24. Ruoslahti E (2012) Peptides as targeting elements and tissue penetration devices for nanoparticles. Adv Mater 24(28):3747–3756
25. Corti A, Pastorino F, Curnis F, Arap W, Ponzoni M, Pasqualini R (2012) Targeted drug delivery and penetration into solid tumors. Med Res Rev 32(5):1078–1091
26. Danhier F, Le Breton A, Preat V (2012) RGD-based strategies to target alpha(v) beta(3) integrin in cancer therapy and diagnosis. Mol Pharm 9(11):2961–2973
27. Kapoor P, Singh H, Gautam A, Chaudhary K, Kumar R, Raghava GPS (2012) TumorHoPe: A database of tumor homing peptides. PLoS One 7(4)
28. Chen K, Chen X (2011) Integrin targeted delivery of Chemotherapeutics. Theranostics 1:189–200
29. Dimitrov I, Trzebicka B, Müller AHE, Dworak A, Tsvetanov CB (2007) Thermosensitive water-soluble copolymers with doubly responsive reversibly interacting entities. Prog Polym Sci 32(11):1275–1343
30. Ramos J, Imaz A, Callejas-Fernández J, Barbosa-Barros L, Estelrich J, Quesada-Pérez M, Forcada J (2011) Soft nanoparticles (thermo-responsive nanogels and bicelles) with biotechnological applications: from synthesis to simulation through colloidal characterization. Soft Matter 7(11):5067
31. Roy D, Brooks WLA, Sumerlin BS (2013) New directions in thermoresponsive polymers. Chem Soc Rev 42(17):7214–7243
32. Talelli M, Rijcken CJF, van Nostrum CF, Storm G, Hennink WE (2010) Micelles based on HPMA copolymers. Adv Drug Deliver Rev 62(2):231–239
33. Schumers J-M, Fustin C-A, Gohy J-F (2010) Light-responsive block copolymers. Macromol Rapid Commun 31(18):1588–1607
34. Zhao Y (2009) Photocontrollable block copolymer micelles: what can we control? J Mater Chem 19(28):4887–4895
35. Husseini GA, Pitt WG (2009) Ultrasonic-activated micellar drug delivery for cancer treatment. J Pharm Sci 98(3):795–811
36. Dai Q, Nelson A (2010) Magnetically-responsive self assembled composites. Chem Soc Rev 39(11):4057
37. Brazel CS (2008) Magnetothermally-responsive nanomaterials: combining magnetic nanostructures and thermally-sensitive polymers for triggered drug release. Pharm Res 26(3):644–656

38. Lee ES, Gao Z, Bae YH (2008) Recent progress in tumor pH targeting nanotechnology. J Control Release 132(3):164–170
39. Kanamala M, Wilson WR, Yang M, Palmer BD, Wu Z (2016) Mechanisms and biomaterials in pH-responsive tumour targeted drug delivery: a review. Biomaterials 85:152–167
40. Park I-K, Singha K, Arote RB, Choi Y-J, Kim WJ, Cho C-S (2010) pH-responsive polymers as gene carriers. Macromol Rapid Commun 31(13):1122–1133
41. Ulbrich K (2004) Polymeric anticancer drugs with pH-controlled activation. Adv Drug Deliver Rev 56(7):1023–1050
42. Gao Y-J, Qiao Z-Y, Wang H (2016) Polymers with tertiary amine groups for drug delivery and bioimaging. Sci China-Chem 59(8):991–1002
43. Hahn ME, Gianneschi NC (2011) Enzyme-directed assembly and manipulation of organic nanomaterials. Chem Commun 47(43):11814–11821
44. Chen Y, Liang G (2012) Enzymatic self-assembly of nanostructures for theranostics. Theranostics 2(2):139–147
45. Hu J, Zhang G, Liu S (2012) Enzyme-responsive polymeric assemblies, nanoparticles and hydrogels. Chem Soc Rev 41(18):5933–5949
46. Zelzer M, Todd SJ, Hirst AR, McDonald TO, Ulijn RV (2013) Enzyme responsive materials: design strategies and future developments. Biomater Sci 1(1):11–39
47. Andresen TL, Thompson DH, Kaasgaard T (2010) Enzyme-triggered nanomedicine: drug release strategies in cancer therapy (invited review). Mol Membr Biol 27(7):353–363
48. Huo M, Yuan J, Tao L, Wei Y (2014) Redox-responsive polymers for drug delivery: from molecular design to applications. Polym Chem 5(5):1519–1528
49. Trachootham D, Alexandre J, Huang P (2009) Targeting cancer cells by ROS-mediated mechanisms: a radical therapeutic approach? Nat Rev Drug Discov 8(7):579–591
50. Song C-C, Du F-S, Li Z-C (2014) Oxidation-responsive polymers for biomedical applications. J Mater Chem B 2(22):3413
51. Meng F, Hennink WE, Zhong Z (2009) Reduction-sensitive polymers and bioconjugates for biomedical applications. Biomaterials 30(12):2180–2198
52. Xie J, Liu G, Eden HS, Ai H, Chen X (2011) Surface-engineered magnetic nanoparticle platforms for cancer imaging and therapy. Acc Chem Res 44(10):883–892
53. Alkilany AM, Lohse SE, Murphy CJ (2013) The gold standard: gold nanoparticle libraries to understand the nano-bio interface. Acc Chem Res 46(3):650–661
54. Tarn D, Ashley CE, Xue M, Carnes EC, Zink JI, Brinker CJ (2013) Mesoporous silica nanoparticle nanocarriers: biofunctionality and biocompatibility. Acc Chem Res 46(3):792–801
55. Sharma A, Sharma US (1997) Liposomes in drug delivery: progress and limitations. Int J Pharm 154(2):123–140
56. Lian T, Ho RJY (2001) Trends and developments in liposome drug delivery systems. J Pharm Sci 90(6):667–680
57. Park JW (2002) Liposome-based drug delivery in breast cancer treatment. Breast cancer research: BCR 4(3):95–99
58. Kraft JC, Freeling JP, Wang Z, Ho RJY (2014) Emerging research and clinical development trends of liposome and lipid nanoparticle drug delivery systems. J Pharm Sci 103(1):29–52
59. van der Meel R, Fens MHAM, Vader P, van Solinge WW, Eniola-Adefeso O, Schiffelers RM (2014) Extracellular vesicles as drug delivery systems: lessons from the liposome field. J Control Release 195:72–85
60. W-d Wu, X-l Yi, L-x Jiang, Y-z Li, Gao J, Zeng Y, R-d Yi, L-p Dai, Li W, X-y Ci, D-y Si, C-x Liu (2015) The targeted-liposome delivery system of antitumor drugs. Curr Drug Metab 16(10):894–910
61. Lavasanifar A, Samuel J, Kwon GS (2002) Poly(ethylene oxide)-block-poly(L-amino acid) micelles for drug delivery. Adv Drug Deliver Rev 54(2):169–190
62. Gaucher G, Dufresne M-H, Sant VP, Kang N, Maysinger D, Leroux J-C (2005) Block copolymer micelles: preparation, characterization and application in drug delivery. J Control Release 109(1–3):169–188

63. Mikhail AS, Allen C (2009) Block copolymer micelles for delivery of cancer therapy: transport at the whole body, tissue and cellular levels. J Control Release 138(3):214–223
64. Meng FH, Zhong ZY, Feijen J (2009) Stimuli-responsive polymersomes for programmed drug delivery. Biomacromol 10(2):197–209
65. Brinkhuis RP, Rutjes FPJT, van Hest JCM (2011) Polymeric vesicles in biomedical applications. Polym Chem 2(7):1449–1462
66. Tanner P, Baumann P, Enea R, Onaca O, Palivan C, Meier W (2011) Polymeric Vesicles: from drug carriers to nanoreactors and artificial organelles. Acc Chem Res 44(10):1039–1049
67. Kabanov AV, Vinogradov SV (2009) Nanogels as pharmaceutical carriers: finite networks of infinite capabilities. Angewandte Chemie-international Edition 48(30):5418–5429
68. Motornov M, Roiter Y, Tokarev I, Minko S (2010) Stimuli-responsive nanoparticles, nanogels and capsules for integrated multifunctional intelligent systems. Prog Polym Sci 35(1–2):174–211
69. Oishi M, Nagasaki Y (2010) Stimuli-responsive smart nanogels for cancer diagnostics and therapy. Nanomedicine 5(3):451–468
70. Gil E, Hudson S (2004) Stimuli-reponsive polymers and their bioconjugates. Prog Polym Sci 29(12):1173–1222
71. Pasut G, Veronese FM (2009) PEG conjugates in clinical development or use as anticancer agents: an overview. Adv Drug Deliver Rev 61(13):1177–1188
72. CdlH Alarcon, Pennadam S, Alexander C (2005) Stimuli responsive polymers for biomedical applications. Chem Soc Rev 34(3):276–285
73. Hoffman AS (2013) Stimuli-responsive polymers: biomedical applications and challenges for clinical translation. Adv Drug Deliver Rev 65(1):10–16
74. Rijcken CJF, Soga O, Hennink WE, van Nostrum CF (2007) Triggered destabilisation of polymeric micelles and vesicles by changing polymers polarity: An attractive tool for drug delivery. J Control Release 120(3):131–148
75. Ge Z, Liu S (2013) Functional block copolymer assemblies responsive to tumor and intracellular microenvironments for site-specific drug delivery and enhanced imaging performance. Chem Soc Rev 42(17):7289–7325
76. Ma X, Tian H (2014) Stimuli-responsive supramolecular polymers in aqueous solution. Acc Chem Res 47(7):1971–1981
77. Siegel RA (2014) Stimuli sensitive polymers and self regulated drug delivery systems: a very partial review. J Control Release 190:337–351
78. Modi S, Swetha MG, Goswami D, Gupta GD, Mayor S, Krishnan Y (2009) A DNA nanomachine that maps spatial and temporal pH changes inside living cells. Nat Nanotechnol 4(5):325–330
79. Casey JR, Grinstein S, Orlowski J (2010) Sensors and regulators of intracellular pH. Nat Rev Mol Cell Biol 11(1):50–61
80. Sahay G, Alakhova DY, Kabanov AV (2010) Endocytosis of nanomedicines. J Control Release 145(3):182–195
81. Iversen T-G, Skotland T, Sandvig K (2011) Endocytosis and intracellular transport of nanoparticles: present knowledge and need for future studies. Nano Today 6(2):176–185
82. Lee ES, Oh KT, Kim D, Youn YS, Bae YH (2007) Tumor pH-responsive flower-like micelles of poly(L-lactic acid)-b-poly (ethylene glycol)-b-poly(L-histidine). J Control Release 123(1):19–26
83. Yin HQ, Lee ES, Kim D, Lee KH, Oh KT, Bae YH (2008) Physicochemical characteristics of pH-sensitive poly(L-Histidine)-b-poly (ethylene glycol)/poly(L-Lactide)-b-poly(ethylene glycol) mixed micelles. J Control Release 126(2):130–138
84. Oh KT, Lee ES, Kim D, Bae YH (2008) L-Histidine-based pH-sensitive anticancer drug carrier micelle: Reconstitution and brief evaluation of its systemic toxicity. Int J Pharm 358(1–2):177–183

85. Kim D, Gao ZG, Lee ES, Bae YH (2009) In vivo evaluation of doxorubicin-loaded polymeric micelles targeting folate receptors and early endosomal pH in drug-resistant ovarian cancer. Mol Pharmaceutics 6(5):1353–1362

86. Lee ES, Gao Z, Kim D, Park K, Kwon IC, Bae YH (2008) Super pH-sensitive multifunctional polymeric micelle for tumor pH(e) specific TAT exposure and multidrug resistance. J Control Release 129(3):228–236

87. Tang YQ, Liu SY, Armes SP, Billingham NC (2003) Solubilization and controlled release of a hydrophobic drug using novel micelle-forming ABC triblock copolymers. Biomacromol 4(6):1636–1645

88. Giacomelli C, Le Men L, Borsali R, Lai-Kee-Him J, Brisson A, Armes SP, Lewis AL (2006) Phosphorylcholine-based pH-responsive diblock copolymer micelles as drug delivery vehicles: light scattering, electron microscopy, and fluorescence experiments. Biomacromol 7(3):817–828

89. Licciardi M, Craparo EF, Giammona G, Armes SP, Tang Y, Lewis AL (2008) In vitro biological evaluation of folate-functionalized block copolymer micelles for selective anti-cancer drug delivery. Macromol Biosci 8 (7):615–626

90. Xu PS, Van Kirk EA, Murdoch WJ, Zhan YH, Isaak DD, Radosz M, Shen YQ (2006) Anticancer efficacies of cisplatin-releasing pH-responsive nanoparticles. Biomacromol 7 (3):829–835

91. Shen Y, Zhan Y, Tang J, Xu P, Johnson PA, Radosz M, Van Kirk EA, Murdoch WJ (2008) Multifunctioning pH-responsive nanoparticles from hierarchical self-assembly of polymer brush for cancer drug delivery. AIChE J 54(11):2979–2989

92. Lynn DM, Langer R (2000) Degradable Poly(β-amino esters): synthesis, characterization, and self-assembly with plasmid DNA. J Am Chem Soc 122(44):10761–10768

93. Akinc A, Anderson DG, Lynn DM, Langer R (2003) Synthesis of poly(β-amino ester)s optimized for highly effective gene delivery. Bioconjug Chem 14(5):979–988

94. Akinc A, Lynn DM, Anderson DG, Langer R (2003) Parallel synthesis and biophysical characterization of a degradable polymer library for gene delivery. J Am Chem Soc 125(18): 5316–5323

95. Anderson DG, Lynn DM, Langer R (2003) Semi-automated synthesis and screening of a large library of degradable cationic polymers for gene delivery. Angew Chem Int Ed 42(27): 3153–3158

96. Green JJ, Langer R, Anderson DG (2008) A combinatorial polymer library approach yields insight into nonviral gene delivery. Acc Chem Res 41(6):749–759

97. Lynn DM, Amiji MM, Langer R (2001) pH-responsive polymer microspheres: rapid release of encapsulated material within the range of intracellular pH. Angew Chem Int Ed 40 (9):1707–1710

98. Shenoy D, Little S, Langer R, Amiji M (2005) Poly(ethylene oxide)-modified poly (beta-amino ester) nanoparticles as a pH-sensitive system for tumor-targeted delivery of hydrophobic drugs. 1. In vitro evaluations. Mol Pharm 2(5):357–366

99. Ko J, Park K, Kim YS, Kim MS, Han JK, Kim K, Park RW, Kim IS, Song HK, Lee DS, Kwon IC (2007) Tumoral acidic extracellular pH targeting of pH-responsive MPEG-poly (beta-amino ester) block copolymer micelles for cancer therapy. J Control Release 123 (2):109–115

100. Min KH, Kim JH, Bae SM, Shin H, Kim MS, Park S, Lee H, Park RW, Kim IS, Kim K, Kwon IC, Jeong SY, Lee DS (2010) Tumoral acidic pH-responsive MPEG-poly(beta-amino ester) polymeric micelles for cancer targeting therapy. J Control Release 144(2):259–266

101. Song W, Tang Z, Li M, Lv S, Yu H, Ma L, Zhuang X, Huang Y, Chen X (2012) Tunable pH-sensitive Poly(beta-amino ester)s synthesized from primary amines and diacrylates for intracellular drug delivery. Macromol Biosci 12(10):1375–1383

102. Qiao Z-Y, Qiao S-L, Fan G, Fan Y-S, Chen Y, Wang H (2014) One-pot synthesis of pH-sensitive poly(RGD-co-[small beta]-amino ester)s for targeted intracellular drug delivery. Polym Chem 5:844–853

103. Qiao Z-Y, Zhang D, Hou C-Y, Zhao S-M, Liu Y, Gao Y-J, Tan N-H, Wang H (2015) A pH-responsive natural cyclopeptide RA-V drug formulation for improved breast cancer therapy. J Mater Chem B 3(22):4514–4523
104. Park SY, Baik HJ, Oh YT, Oh KT, Youn YS, Lee ES (2011) A smart polysaccharide/drug conjugate for photodynamic therapy. Angew Chem Int Ed 50(7):1644–1647
105. Qiao Z-Y, Hou C-Y, Zhang D, Liu Y, Lin Y-X, An H-W, Li X-J, Wang H (2015) Self-assembly of cytotoxic peptide conjugated Poly([small beta]-amino ester)s for synergistic cancer chemotherapy. J Mater Chem B 3:2943–2953
106. Wang Y, Lin Y-X, Qiao Z-Y, An H-W, Qiao S-L, Wang L, Rajapaksha RPYJ, Wang H (2015) Self-assembled autophagy-inducing polymeric nanoparticles for breast cancer interference in-vivo. Adv Mater 27(16):2627–2634
107. Gillies ER, Jonsson TB, Frechet JMJ (2004) Stimuli-responsive supramolecular assemblies of linear-dendritic copolymers. J Am Chem Soc 126(38):11936–11943
108. Gillies ER, Frechet JMJ (2005) pH-responsive copolymer assemblies for controlled release of doxorubicin. Bioconjug Chem 16(2):361–368
109. Gillies ER, Frechet JMJ (2003) A new approach towards acid sensitive copolymer micelles for drug delivery. Chem Commun 14:1640–1641
110. Jain R, Standley SM, Frechet JMJ (2007) Synthesis and degradation of pH-sensitive linear poly(amidoamine)s. Macromolecules 40(3):452–457
111. Bachelder EM, Beaudette TT, Broaders KE, Paramonov SE, Dashe J, Frechet JMJ (2008) Acid-degradable polyurethane particles for protein-based vaccines: Biological evaluation and in vitro analysis of particle degradation products. Mol Pharm 5(5):876–884
112. Heffernan MJ, Murthy N (2005) Polyketal nanoparticles: A new pH-sensitive biodegradable drug delivery vehicle. Bioconjug Chem 16(6):1340–1342
113. Yang SC, Bhide M, Crispe IN, Pierce RH, Murthy N (2008) Polyketal copolymers: A new acid-sensitive delivery vehicle for treating acute inflammatory diseases. Bioconjug Chem 19 (6):1164–1169
114. Chen W, Meng FH, Li F, Ji SJ, Zhong ZY (2009) pH-responsive biodegradable micelles based on acid-labile polycarbonate hydrophobe: synthesis and triggered drug release. Biomacromol 10(7):1727–1735
115. Chen W, Meng FH, Cheng R, Zhong ZY (2010) pH-Sensitive degradable polymersomes for triggered release of anticancer drugs: a comparative study with micelles. J Control Release 142(1):40–46
116. Griset AP, Walpole J, Liu R, Gaffey A, Colson YL, Grinstaff MW (2009) Expansile nanoparticles: synthesis, characterization, and in vivo efficacy of an acid-responsive polymeric drug delivery system. J Am Chem Soc 131(7):2469–2471
117. Shim MS, Kwon YJ (2008) Controlled delivery of plasmid DNA and siRNA to intracellular targets using ketalized polyethylenimine. Biomacromol 9(2):444–455
118. Shim MS, Kwon YJ (2009) Controlled cytoplasmic and nuclear localization of plasmid DNA and siRNA by differentially tailored polyethylenimine. J Control Release 133(3): 206–213
119. Bachelder EM, Beaudette TT, Broaders KE, Dashe J, Frechet JMJ (2008) Acetal-derivatized dextran: an acid-responsive biodegradable material for therapeutic applications. J Am Chem Soc 130(32):10494–10495
120. Broaders KE, Cohen JA, Beaudette TT, Bachelder EM, Frechet JMJ (2009) Acetalated dextran is a chemically and biologically tunable material for particulate immunotherapy. Proc Natl Acad Sci USA 106(14):5497–5502
121. Cohen JA, Beaudette TT, Cohen JL, Brooders KE, Bachelder EM, Frechet JMJ (2010) Acetal-modified dextran microparticles with controlled degradation kinetics and surface functionality for gene delivery in phagocytic and non-phagocytic cells. Adv Mater 22 (32): 3593
122. Heller J, Barr J (2004) Poly(ortho esters) - From concept to reality. Biomacromol 5(5): 1625–1632

123. Wang C, Ge Q, Ting D, Nguyen D, Shen HR, Chen JZ, Eisen HN, Heller J, Langer R, Putnam D (2004) Molecularly engineered poly(ortho ester) microspheres for enhanced delivery of DNA vaccines. Nat Mater 3(3):190–196

124. Qiao Z-Y, Du FS, Zhang R, Liang DH, Li ZC (2010) Biocompatible thermoresponsive polymers with pendent Oligo(ethylene glycol) Chains and cyclic ortho ester groups. Macromolecules 43(15):6485–6494

125. Qiao Z-Y, Cheng J, Ji R, Du F-S, Liang D-H, Ji S-P, Li Z-C (2013) Biocompatible acid-labile polymersomes from PEO-b-PVA derived amphiphilic block copolymers. RSC Adv 3(46):24345–24353

126. Huang XN, Du FS, Ju R, Li ZC (2007) Novel acid-labile, thermoresponsive poly (methacrylamide)s with pendent ortho ester moieties. Macromol Rapid Commun 28(5): 597–603

127. Huang XN, Du FS, Cheng J, Dong YQ, Liang DH, Ji SP, Lin SS, Li ZC (2009) Acid-sensitive polymeric micelles based on thermoresponsive block copolymers with pendent cyclic orthoester groups. Macromolecules 42(3):783–790

128. Qiao Z-Y, Ji R, Huang X-N, Du F-S, Zhang R, Liang D-H, Li Z-C (2013) Polymersomes from dual responsive block copolymers: drug encapsulation by heating and acid-triggered release. Biomacromol 14(5):1555–1563

129. Tang R, Ji W, Panus D, Palumbo RN, Wang C (2011) Block copolymer micelles with acid-labile ortho ester side-chains: synthesis, characterization, and enhanced drug delivery to human glioma cells. J Control Release 151(1).18–27

130. Hruby M, Konak C, Ulbrich K (2005) Polymeric micellar pH-sensitive drug delivery system for doxorubicin. J Control Release 103(1):137–148

131. Chytil P, Etrych T, Konak C, Sirova M, Mrkvan T, Rihova B, Ulbrich K (2006) Properties of HPMA copolymer-doxorubicin conjugates with pH-controlled activation: Effect of polymer chain modification. J Control Release 115(1):26–36

132. Etrych T, Chytil P, Mrkvan T, Sirova M, Rihova B, Ulbrich K (2008) Conjugates of doxorubicin with graft HPMA copolymers for passive tumor targeting. J Control Release 132(3):184–192

133. Etrych T, Mrkvan T, Rihova B, Ulbrich K (2007) Star-shaped immunoglobulin-containing HPMA-based conjugates with doxorubicin for cancer therapy. J Control Release 122(1): 31–38

134. Etrych T, Sirova M, Starovoytova L, Rihova B, Ulbrich K (2010) HPMA copolymer conjugates of paclitaxel and docetaxel with ph-controlled drug release. Mol Pharm 7(4): 1015–1026

135. Chytil P, Koziolova E, Janouskova O, Kostka L, Ulbrich K, Etrych T (2015) Synthesis and properties of Star HPMA copolymer nanocarriers synthesised by raft polymerisation designed for selective anticancer drug delivery and imaging. Macromol Biosci 15(6):839–850

136. Dozono H, Yanazume S, Nakamura H, Etrych T, Chytil P, Ulbrich K, Fang J, Arimura T, Douchi T, Kobayashi H, Ikoma M, Maeda H (2016) HPMA copolymer-conjugated pirarubicin in multimodal treatment of a patient with stage IV prostate cancer and extensive lung and bone metastases. Target Oncol 11(1):101–106

137. Kim D, Lee ES, Park K, Kwon IC, Bae YH (2008) Doxorubicin loaded pH-sensitive micelle: antitumoral efficacy against ovarian A2780/DOXR tumor. Pharm Res 25(9):2074–2082

138. Bae Y, Nishiyama N, Fukushima S, Koyama H, Yasuhiro M, Kataoka K (2005) Preparation and biological characterization of polymeric micelle drug carriers with intracellular pH-triggered drug release property: Tumor permeability, controlled subcellular drug distribution, and enhanced in vivo antitumor efficacy. Bioconjug Chem 16(1):122–130

139. Prabaharan M, Grailer JJ, Pilla S, Steeber DA, Gong SQ (2009) Amphiphilic multi-arm-block copolymer conjugated with doxorubicin via pH-sensitive hydrazone bond for tumor-targeted drug delivery. Biomaterials 30(29):5757–5766

140. Wang WL, Wang B, Ma XJ, Liu SR, Shang XD, Yu XF (2016) Tailor-made pH-Responsive Poly(choline phosphate) prodrug as a drug delivery system for rapid cellular internalization. Biomacromol 17(6):2223–2232

141. Jia ZF, Wong LJ, Davis TP, Bulmus V (2008) One-Pot conversion of RAFT-generated multifunctional block copolymers of HPMA to doxorubicin conjugated acid- and reductant-sensitive crosslinked micelles. Biomacromol 9(11):3106–3113

142. Wang Y, Zhang L, Zhang XB, Wei X, Tang ZM, Zhou SB (2016) Precise polymerization of a highly tumor microenvironment-responsive nanoplatform for strongly enhanced intracellular drug release. ACS Appl Mater Interfaces 8(9):5833–5846

143. Zhang Y, Ding JX, Li MQ, Chen X, Xiao CS, Zhuang XL, Huang YB, Chen XS (2016) One-step "click chemistry"-synthesized cross-linked prodrug nanogel for highly selective intracellular drug delivery and upregulated antitumor efficacy. ACS Appl Mater Interfaces 8 (17):10673–10682

144. Lu H, Wang DL, Kazane S, Javahishvili T, Tian F, Song F, Sellers A, Barnett B, Schultz PG (2013) Site-specific antibody-polymer conjugates for siRNA delivery. J Am Chem Soc 135 (37):13885–13891

145. Yoo HS, Lee EA, Park TG (2002) Doxorubicin-conjugated biodegradable polymeric micelles having acid-cleavable linkages. J Control Release 82(1):17–27

146. Zhu L, Zhao L, Qu X, Yang Z (2012) pH-sensitive polymeric vesicles from coassembly of amphiphilic cholate grafted poly(l-lysine) and acid-cleavable polymer-drug conjugate. Langmuir 28(33):11988–11996

147. Yang S, Zhu F, Wang Q, Liang F, Qu X, Gan Z, Yang Z (2015) Combinatorial targeting polymeric micelles for anti-tumor drug delivery. J Mater Chem B 3(19):4043–4051

148. Sawant RM, Hurley JP, Salmaso S, Kale A, Tolcheva E, Levchenko TS, Torchilin VP (2006) "SMART" drug delivery systems: double-targeted pH-responsive pharmaceutical nanocarriers. Bioconjug Chem 17(4):943–949

149. Ding CX, Gu JX, Qu XZ, Yang ZZ (2009) Preparation of multifunctional drug carrier for tumor-specific uptake and enhanced intracellular delivery through the conjugation of weak acid labile linker. Bioconjug Chem 20(6):1163–1170

150. Van Horn BA, Iha RK, Wooley KL (2008) Sequential and single-step, one-pot strategies for the transformation of hydrolytically degradable polyesters into multifunctional systems. Macromolecules 41(5):1618–1626

151. Carmona S, Jorgensen MR, Kolli S, Crowther C, Salazar FH, Marion PL, Fujino M, Natori Y, Thanou M, Arbuthnot P, Miller AD (2009) Controlling HBV replication in vivo by intravenous administration of triggered PEGylated siRNA-nanoparticles. Mol Pharm 6 (3):706–717

152. Kolli S, Wong S-P, Harbottle R, Johnston B, Thanou M, Miller AD (2013) pH-triggered nanoparticle mediated delivery of siRNA to liver cells in vitro and in vivo. Bioconj Chem 24 (3):314–332

153. Oishi M, Sasaki S, Nagasaki Y, Kataoka K (2003) PH-Responsive oligodeoxynucleotide (ODN)-poly(ethylene glycol) conjugate through acid-labile beta-thiopropionate linkage: Preparation and polyion complex micelle formation. Biomacromol 4(5):1426–1432

154. Oishi M, Nagasaki Y, Itaka K, Nishiyama N, Kataoka K (2005) Lactosylated poly(ethylene glycol)-siRNA conjugate through acid-labile ss-thiopropionate linkage to construct pH-sensitive polyion complex micelles achieving enhanced gene silencing in hepatoma cells. J Am Chem Soc 127(6):1624–1625

155. Dan K, Ghosh S (2013) One-pot synthesis of an acid-labile amphiphilic triblock copolymer and its ph-responsive vesicular assembly. Angew Chem Int Ed 52(28):7300–7305

156. Rozema DB, Lewis DL, Wakefield DH, Wong SC, Klein JJ, Roesch PL, Bertin SL, Reppen TW, Chu Q, Blokhin AV, Hagstrom JE, Wolff JA (2007) Dynamic PolyConjugates for targeted in vivo delivery of siRNA to hepatocytes. Proc Natl Acad Sci USA 104 (32):12982–12987

157. Lee Y, Miyata K, Oba M, Ishii T, Fukushima S, Han M, Koyama H, Nishiyama N, Kataoka K (2008) Charge-conversion ternary polyplex with endosome disruption moiety: a

technique for efficient and safe gene delivery. Angewandte Chemie-Int Edition 47(28):5163–5166

158. Lee Y, Ishii T, Cabral H, Kim HJ, Seo JH, Nishiyama N, Oshima H, Osada K, Kataoka K (2009) Charge-conversional polyionic complex micelles-efficient nanocarriers for protein delivery into cytoplasm. Angewandte Chemie-International Edition 48(29):5309–5312

159. Takemoto H, Miyata K, Hattori S, Ishii T, Suma T, Uchida S, Nishiyama N, Kataoka K (2013) Acidic pH-responsive siRNA conjugate for reversible carrier stability and accelerated endosomal escape with reduced IFN alpha-associated immune response. Angewandte Chemie-Int Edition 52(24):6218–6221

160. Hu FQ, Zhang YY, You J, Yuan H, Du YZ (2012) pH triggered doxorubicin delivery of PEGylated glycolipid conjugate micelles for tumor targeting therapy. Mol Pharm 9(9):2469–2478

161. Lavignac N, Nicholls JL, Ferruti P, Duncan R (2009) Poly(amidoamine) conjugates containing doxorubicin bound via an acid-sensitive linker. Macromol Biosci 9(5):480–487

162. Yuan L, Chen WL, Li J, Hu JH, Yan JJ, Yang D (2012) PEG-b-PtBA-b-PHEMA well-defined amphiphilic triblock copolymer: Synthesis, self-assembly, and application in drug delivery. J Polymer Sci Part Polymer Chem 50(21):4579–4588

163. Wang K, Zhang XF, Liu Y, Liu C, Jiang BH, Jiang YY (2014) Tumor penetrability and anti-angiogencsis using iRGD-mediated delivery of doxorubicin-polymer conjugates. Biomaterials 35(30):8735–8747

164. Lin CJ, Kuan CH, Wang LW, Wu HC, Chen YC, Chang CW, Huang RY, Wang TW (2016) Integrated self-assembling drug delivery system possessing dual responsive and active targeting for orthotopic ovarian cancer theranostics. Biomaterials 90:12–26

165. Maier K, Wagner E (2012) Acid-labile traceless click linker for protein transduction. J Am Chem Soc 134(24):10169–10173

166. Li HJ, Du JZ, Du XJ, Xu CF, Sun CY, Wang HX, Cao ZT, Yang XZ, Zhu YH, Nie SM, Wang J (2016) Stimuli-responsive clustered nanoparticles for improved tumor penetration and therapeutic efficacy. Proc Natl Acad Sci USA 113(15):4164–4169

167. Sun HL, Guo BN, Li XQ, Cheng R, Meng FH, Liu HY, Zhong ZY (2010) Shell-sheddable micelles based on dextran-SS-Poly(epsilon-caprolactone) Diblock copolymer for efficient intracellular release of doxorubicin. Biomacromol 11(4):848–854

168. Zhong YN, Yang WJ, Sun HL, Cheng R, Meng FH, Deng C, Zhong ZY (2013) Ligand-directed reduction-sensitive shell-sheddable biodegradable micelles actively deliver doxorubicin into the nuclei of target cancer cells. Biomacromol 14(10):3723–3730

169. Zhu YQ, Zhang J, Meng FH, Deng C, Cheng R, Jan FJ, Zhong ZY (2016) cRGD-functionalized reduction-sensitive shell-sheddable biodegradable micelles mediate enhanced doxorubicin delivery to human glioma xenografts in vivo. J Control Release 233:29–38

170. Wang XX, Zhang J, Cheng R, Meng FH, Deng C, Zhong ZY (2016) Facile synthesis of reductively degradable biopolymers using cystamine diisocyanate as a coupling agent. Biomacromol 17(3):882–890

171. Chen W, Zhong P, Meng FH, Cheng R, Deng C, Feijen J, Zhong ZY (2013) Redox and pH-responsive degradable micelles for dually activated intracellular anticancer drug release. J Control Release 169(3):171–179

172. Klaikherd A, Nagamani C, Thayumanavan S (2009) Multi-stimuli sensitive amphiphilic block copolymer assemblies. J Am Chem Soc 131(13):4830–4838

173. Qiao Z-Y, Zhang R, Du FS, Liang DH, Li ZC (2011) Multi-responsive nanogels containing motifs of ortho ester, oligo(ethylene glycol) and disulfide linkage as carriers of hydrophobic anti-cancer drugs. J Control Release 152(1):57–66

174. Oh JK, Siegwart DJ, Lee HI, Sherwood G, Peteanu L, Hollinger JO, Kataoka K, Matyjaszewski K (2007) Biodegradable nanogels prepared by atom transfer radical polymerization as potential drug delivery carriers: Synthesis, biodegradation, in vitro release, and bioconjugation. J Am Chem Soc 129(18):5939–5945

175. Oh JK, Tang CB, Gao HF, Tsarevsky NV, Matyjaszewski K (2006) Inverse miniemulsion ATRP: A new method for synthesis and functionalization of well-defined water-soluble/cross-linked polymeric particles. J Am Chem Soc 128(16):5578–5584
176. Ja-Hyoung Ryu RTC, Siriporn Jiwpanich, Sean Bickerton, R. Prakash Babu, S. Thayumanavan (2010) Self-cross-linked polymer nanogels: a versatile nanoscopic drug delivery platform. J Am Chem Soc 132:17227–17235
177. Dai J, Lin SD, Cheng D, Zou SY, Shuai XT (2011) Interlayer-crosslinked micelle with partially hydrated core showing reduction and pH dual sensitivity for pinpointed intracellular drug release. Angew Chem Int Ed 50(40):9404–9408
178. Chen W, Zheng M, Meng FH, Cheng R, Deng C, Feijen J, Zhong ZY (2013) In situ forming reduction-sensitive degradable nanogels for facile loading and triggered intracellular release of proteins. Biomacromol 14(4):1214–1222
179. Wu LL, Zou Y, Deng C, Cheng R, Meng FH, Zhong ZY (2013) Intracellular release of doxorubicin from core-crosslinked polypeptide micelles triggered by both pH and reduction conditions. Biomaterials 34(21):5262–5272
180. Chen W, Meng FH, Cheng R, Deng C, Feijen J, Zhong Z (2015) Facile construction of dual-bioresponsive biodegradable micelles with superior extracellular stability and activated intracellular drug release. J Control Release 210:125–133
181. MacKay CE, Knock GA (2015) Control of vascular smooth muscle function by Src-family kinases and reactive oxygen species in health and disease. J Physiol London 593(17):3815–3828
182. Panieri E, Santoro MM (2015) ROS signaling and redox biology in endothelial cells. Cell Mol Life Sci 72(17):3281–3303
183. Burgoyne JR, S-i Oka, Ale-Agha N, Eaton P (2012) Hydrogen peroxide sensing and signaling by protein kinases in the cardiovascular system. Antioxid Redox Signal 18 (9):1042–1052
184. Chen X, Tian X, Shin I, Yoon J (2011) Fluorescent and luminescent probes for detection of reactive oxygen and nitrogen species. Chem Soc Rev 40(9):4783–4804
185. Xu Q, He C, Xiao C, Chen X (2016) Reactive oxygen species (ROS) responsive polymers for biomedical applications. Macromol Biosci 16(5):635–646
186. Joshi-Barr S, de Gracia Lux C, Mahmoud E, Almutairi A (2013) Exploiting oxidative microenvironments in the body as triggers for drug delivery systems. Antioxid Redox Signal 21(5):730–754
187. Li X, Makarov SS (2006) An essential role of NF-κB in the "tumor-like" phenotype of arthritic synoviocytes. Proc Natl Acad Sci USA 103(46):17432–17437
188. Schäfer M, Werner S (2008) Oxidative stress in normal and impaired wound repair. Pharmacol Res 58(2):165–171
189. Napoli A, Valentini M, Tirelli N, Muller M, Hubbell JA (2004) Oxidation-responsive polymeric vesicles. Nat Mater 3(3):183–189
190. Liu B, Wang D, Liu Y, Zhang Q, Meng L, Chi H, Shi J, Li G, Li J, Zhu X (2015) Hydrogen peroxide-responsive anticancer hyperbranched polymer micelles for enhanced cell apoptosis. Polym Chem 6(18):3460–3471
191. Gupta MK, Meyer TA, Nelson CE, Duvall CL (2012) Poly(PS-b-DMA) micelles for reactive oxygen species triggered drug release. J Control Release 162(3):591–598
192. Poole KM, Nelson CE, Joshi RV, Martin JR, Gupta MK, Haws SC, Kavanaugh TE, Skala MC, Duvall CL (2015) ROS-responsive microspheres for on demand antioxidant therapy in a model of diabetic peripheral arterial disease. Biomaterials 41:166–175
193. Qiao Z-Y, Zhao W-J, Cong Y, Zhang D, Hu Z, Duan Z-Y, Wang H (2016) Self-assembled ros-sensitive polymer-peptide therapeutics incorporating built-in reporters for evaluation of treatment efficacy. Biomacromol 17(5):1643–1652
194. Xiao C, Ding J, Ma L, Yang C, Zhuang X, Chen X (2015) Synthesis of thermal and oxidation dual responsive polymers for reactive oxygen species (ROS)-triggered drug release. Polym Chem 6(5):738–747

195. Zhao WJ, Qiao Z-Y, Duan ZY, Wang H (2016) Synthesis and self-assembly of pH and ROS dual responsive Poly(beta-thioester)s. Acta Chim Sinica 74(3):234–240
196. Xu H, Cao W, Zhang X (2013) Selenium-containing polymers: promising biomaterials for controlled release and enzyme mimics. Acc Chem Res 46(7):1647–1658
197. Ma N, Li Y, Ren H, Xu H, Li Z, Zhang X (2010) Selenium-containing block copolymers and their oxidation-responsive aggregates. Polym Chem 1(10):1609–1614
198. Cao W, Li Y, Yi Y, Ji S, Zeng L, Sun Z, Xu H (2012) Coordination-responsive selenium-containing polymer micelles for controlled drug release. Chem Sci 3(12):3403–3408
199. Liu J, Pang Y, Zhu Z, Wang D, Li C, Huang W, Zhu X, Yan D (2013) Therapeutic nanocarriers with hydrogen peroxide-triggered drug release for cancer treatment. Biomacromol 14(5):1627–1636
200. Ma N, Li Y, Xu H, Wang Z, Zhang X (2010) Dual redox responsive assemblies formed from diselenide block copolymers. J Am Chem Soc 132(2):442–443
201. Roy D, Cambre JN, Sumerlin BS (2010) Future perspectives and recent advances in stimuli-responsive materials. Prog Polym Sci 35(1–2):278–301
202. Broaders KE, Grandhe S, Fréchet JMJ (2011) A biocompatible oxidation-triggered carrier polymer with potential in therapeutics. J Am Chem Soc 133(4):756–758
203. Wang M, Sun S, Neufeld CI, Perez-Ramirez B, Xu Q (2014) Reactive oxygen species-responsive protein modification and its intracellular delivery for targeted cancer therapy. Angew Chem Int Ed 53(49):13444–13448
204. Deng Z, Qian Y, Yu Y, Liu G, Hu J, Zhang G, Liu S (2016) Engineering intracellular delivery nanocarriers and nanoreactors from oxidation-responsive polymersomes via synchronized bilayer cross-linking and permeabilizing inside live cells. J Am Chem Soc
205. Zhang M, Song C-C, Ji R, Qiao Z-Y, Yang C, Qiu F-Y, Liang D-H, Du F-S, Li Z C (2016) Oxidation and temperature dual responsive polymers based on phenylboronic acid and N-isopropylacrylamide motifs. Polym Chem 7(7):1494–1504
206. Nicolaou KC, Mathison CJN, Montagnon T (2003) New reactions of IBX: oxidation of nitrogen- and sulfur-containing substrates to afford useful synthetic intermediates. Angew Chem Int Ed 42(34):4077–4082
207. Shim MS, Xia Y (2013) A Reactive Oxygen Species (ROS)-Responsive Polymer for Safe, Efficient, and Targeted Gene Delivery in Cancer Cells. Angew Chem Int Ed 52(27):6926–6929
208. Yuan YY, Liu J, Liu B (2014) Conjugated-polyelectrolyte-based polyprodrug: targeted and image-guided photodynamic and chemotherapy with on-demand drug release upon irradiation with a single light source. Angew Chem Int Ed 53(28):7163–7168
209. Rauhut MM (1969) Chemiluminescence from concerted peroxide decomposition reactions. Acc Chem Res 2(3):80–87
210. Rauhut MM, Bollyky LJ, Roberts BG, Loy M, Whitman RH, Iannotta AV, Semsel AM, Clarke RA (1967) Chemiluminescence from reactions of electronegatively substituted aryl oxalates with hydrogen peroxide and fluorescent compounds. J Am Chem Soc 89(25):6515–6522
211. Lee D, Khaja S, Velasquez-Castano JC, Dasari M, Sun C, Petros J, Taylor WR, Murthy N (2007) In vivo imaging of hydrogen peroxide with chemiluminescent nanoparticles, vol 6, 10
212. de Gracia Lux C, Joshi-Barr S, Nguyen T, Mahmoud E, Schopf E, Fomina N, Almutairi A (2012) Biocompatible polymeric nanoparticles degrade and release cargo in response to biologically relevant levels of hydrogen peroxide. J Am Chem Soc 134(38):15758–15764
213. Shuhendler AJ, Pu K, Cui L, Uetrecht JP, Rao J (2014) Real-time imaging of oxidative and nitrosative stress in the liver of live animals for drug-toxicity testing. Nat Biotech 32(4):373–380
214. Lee Y-D, Lim C-K, Singh A, Koh J, Kim J, Kwon IC, Kim S (2012) Dye/peroxalate aggregated nanoparticles with enhanced and tunable chemiluminescence for biomedical imaging of hydrogen peroxide. ACS Nano 6(8):6759–6766

215. Lee D, Bae S, Hong D, Lim H, Yoon JH, Hwang O, Park S, Ke Q, Khang G, Kang PM (2013) H$_2$O$_2$-responsive molecularly engineered polymer nanoparticles as ischemia/reperfusion-targeted nanotherapeutic agents. Sci Rep 3:2233
216. Lee SH, Boire TC, Lee JB, Gupta MK, Zachman AL, Rath R, Sung H-J (2014) ROS-cleavable proline oligomer crosslinking of polycaprolactone for pro-angiogenic host response. J Mat Chem B 2(41):7109–7113

# Chapter 6
# Thermoresponsive Polymeric Assemblies and Their Biological Applications

**Sheng-Lin Qiao and Hao Wang**

**Abstract** Thermoresponsive polymeric assemblies are fabricated from thermoresponsive polymers based on the soluble to insoluble transition at the phase transition temperature. The architecture of the self-assemblies lies in the molecular level and influenced by the structural change of the thermoresponsive polymers. The formed nanoassemblies endow excellent biofunctions for biomedical applications. Typical thermoresponsive polymers are reviewed in this chapter. The design concept and key factors to influence the thermal behavior are also discussed and summarized. As the sprouting and promising field, in vivo self-assembled thermoresponsive polymeric assemblies are also discussed in this chapter.

## 6.1 Introduction

Thermoresponsive materials are flourishing and have triggered a variety of useful assemblies that for biomedical applications since the earliest report of the thermal phase transition behavior of poly(N-isopropylacrylamide) (PNIPAAm) in 1967 by Scarpa et al. [1] There are two types of thermoresponsive materials, i.e., lower critical solution temperature (LCST) type and upper critical solution temperature (UCST) type. As (co)polymers with UCST feature are relatively less studied that limit their applications, this chapter will focus on temperature-responsive copolymers that have LCST behavior. LCST is the temperature below which the materials

S.-L. Qiao · H. Wang (✉)
CAS Center for Excellence in Nanoscience, CAS Key Laboratory for Biomedical
Effects of Nanomaterials and Nanosafety, National Center for Nanoscience and
Technology (NCNST), Beijing 100190, People's Republic of China
e-mail: wanghao@nanoctr.cn

S.-L. Qiao
e-mail: qiaosl@nanoctr.cn

S.-L. Qiao · H. Wang
University of Chinese Academy of Sciences (UCAS), Beijing 100049,
People's Republic of China

© Springer Nature Singapore Pte Ltd. 2018
H. Wang and L.-L. Li (eds.), *In Vivo Self-Assembly Nanotechnology for Biomedical Applications*, Nanomedicine and Nanotoxicology,
https://doi.org/10.1007/978-981-10-6913-0_6

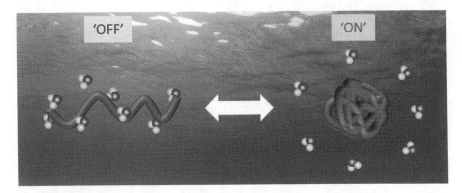

**Fig. 6.1** Schematic illustration of the phase transition behavior of thermoresponsive polymers in aqueous solution

completely solvate in aqueous phase and become insoluble and phase-separated above this temperature (Fig. 6.1). This process is reversible. The phase transition is an energy-driven process which depends on the free energy of mixing or the entropy of the system [2].

Several studies have highlighted the abnormal temperatures in tumors and other inflammatory diseases as a direct result of abnormal blood flow, leukocyte infiltration, a high rate of metabolic activity, and a high rate of cell proliferation in diseased tissues [3]. Besides these intrinsic temperature variations, larger temperature changes can be induced artificially at specific locations by applying heat from an external source, which exploits the higher sensitivity of tumor tissues to high temperatures as compared to normal tissues. Both intrinsic tumor temperature variations and externally induced hyperthermic temperature changes offer attractive stimuli for the site specific controlling the assembly or disassembly of the thermoresponsive materials.

Naturally occurring thermoresponsive polymers represent a large class. Especially, elastin-like polypeptide (ELP), whose phase behavior is encoded in its molecular level, is the most promising ones in this category. By carefully designing, different architecture of ELPs and resulting assemblies are obtained, which can meet the needs of different biological applications. Despite highly biodegradable and biocompatible in nature, the batch-to-batch variability and broad molecular weight (MW) distributions of most of the natural polymers make them less attractive than synthetic polymers that are more reproducible and versatile. Poly (N-isopropylacrylamide) (PNIPAAm) is the leading member of the synthetic thermoresponsive polymers due to its sharp phase transition at 32 °C (a temperature near physiological temperature), and minimal dependent on concentration and molecular weight. To render more intriguing properties, some of the smart nanosystems are not only composed of thermoresponsive units, but also of other building blocks that have a structural function acting as triggers or labels.

This chapter focuses on the thermoresponsive material-regulated assemblies from the viewpoints of their molecular design, self-assembly mechanism, formed structures, and biofunctions. In the following sections of this chapter, we review a variety of thermoresponsive polymeric assemblies used for biomedical applications. The chapter ends with an outlook of some of the future trends in biotechnology and biomedicine applications.

## 6.2 Naturally Occurring Thermoresponsive Polymeric Nanoassemblies and Their Bioapplications

Naturally occurring biodegradable thermoresponsive polymers offer great opportunities for biomedical applications, such as imaging, dug/protein delivery, tissue engineering, and bioscaffold design, due to their abundance in nature and biocompatibility. These materials include protein-based polymers, such as elastin, collagen, and polysaccharides, such as cellulose and cellulose derivatives, chitosan, and chitosan derivatives.

### 6.2.1 Elastin-Like Polypeptides (ELPs) Regulated Assembly

In the past few decades, thermoresponsive nanostructures produced with peptide domains from the extracellular matrix have been intensely studied to develop new smart materials such as hydrogels, films, and drug nanocarriers. Canonized elastin-like polypeptides (ELPs) that derived from the elastometic domain of mammalian tropoelastin [4] and comprise repetitions of the pentapeptide sequence Val-Pro-Gly-Xaa-Gly (where the "guest residue," Xaa, can be any mixture of amino acids except proline), have been very widely studied owing to their lower critical solution temperature (LCST)-type behavior. They are highly soluble in water below the phase transition temperature surrounding by clatharate-like water structures, [5] but the chains fold and assemble into a dynamic, regular, nonrandom structure known as a β-spiral when the temperature is raised above their transition temperatures. This structure contains type II β-turns [6] as the main secondary feature, which is formed by intrachain hydrogen bonding around the L-proline residue and is stabilized by interturn. In this arrangement, the apolar side chains of the constituting amino acids point to the outer shell of the β-spiral, facilitating the existence of extensive intramolecular hydrophobic contacts and, finally, phase separation. The phase transition of these protein-based polymers can be described in terms of an increase in order. It occurs due to hydrophobic folding and assembly.

The LCST of an ELP can be tuned at the molecular level by guest residue composition and distribution, by molecular weight (MW) of the ELP, and by solution parameters such as pH, ionic strength, and ELP concentration. Molecular biology strategy also provides control over the type, number, and location of reactive sites along the polymer chain needed for chemical conjugation of extrinsic moieties to the ELP or cross-linking of these polypeptides. Urry demonstrated the LCST of an ELP is inversely related to the hydrophobicity of the guest residues and can be adjusted over a wide range of temperatures [7]. Incorporation of the hydrophobic amino acid isoleucine rather than valine at the guest position reduced the LCST to below ambient temperatures, whereas incorporation of more hydrophilic lysine raised the LCST from 12–27 °C without affecting the narrow temperature range (2–3 °C) of the response. This effect is proportional to the mole fraction of a specific guest residue. Besides, for canonized ELPs, the phase behavior of a polypeptide is encoded in its amino acid sequence. Chilkoti group [8] synthesized a series of intrinsically disordered, repeat polypeptides and demonstrated that the polypeptides can be designed to exhibit tunable LCST or UCST behaviors in physiological solutions. They also showed that mutation of key residues at the repeat level abolishes phase behavior or encodes an orthogonal transition.

Self-assemblies with diverse biomedical applications can be obtained based on elastin-like block co-recombinamers. Chilkoti group first developed materials based on the canonized elastin pentapeptide repeat Val-Pro-Gly-Xaa-Gly (where the guest residue Xaa = Val/Ala/Gly [1:8:7]), and engineered to exhibit LCST behavior around 40 °C. The transition temperatures of these materials were designed such that near-monodisperse, sub-100-nm-sized nanoparticles might form upon ultrasound induction of hyperthermia as in this way local targeting of drugs conjugated to an ELP backbone might be possible. The thermal transitions for homopolypeptides occurred over a narrow range and were fully reversible for a $(Val_5-Ala_2-Gly_3)_{150}$ polypeptide, and the LCST onset was at 40 °C and complete by 42 °C. Block copolymers were also prepared, but the thermal transitions were more complex indicating a range of intermediate species formed as differential blocks aggregated. For both homo- and copolymers, particles of 40–100 nm were produced above the LCST suggesting advantageous use in cancer therapies owing to the accumulation of particles of this size in tumor tissues. Recently, the same group has reported the use of ELP–doxorubicin conjugates for temperature-mediated tumor suppression [9]. The conjugates were taken into squamous cell carcinoma cells (FaDu) and trafficked into lysosomes via endocytosis, but there were no differences in the in vitro cytotoxicity of free doxorubicin and the ELP–doxorubicin conjugates, even though their subcellular localization was significantly different. Further work is undoubtedly necessary to optimize these materials; however, the local accumulation of the ELP conjugates in tumor cells is a promising first step in drug targeting by responsive synthetic polypeptides.

Short ELPs (e.g., those with fewer than 10 pentapeptides) are rarely used for the thermoresponsive fabrication of nanoparticles, owing to their high transition temperatures. However, recently, Kiick and Luo utilized short ELP with the sequence $(VPGFG)_6$ to fabricate thermoresponsive vesicles (Fig. 6.2) [10]. They conjugated the $(VPGFG)_6$ to a hydrophilic CLP $((GPO)_4GFOGER(GPO)_4GG)$, which resulted in a dramatic reduction of the transition temperature of the ELP by 80 °C owing to the unexpected anchoring effects of the CLP triple helix in the ELP−CLP

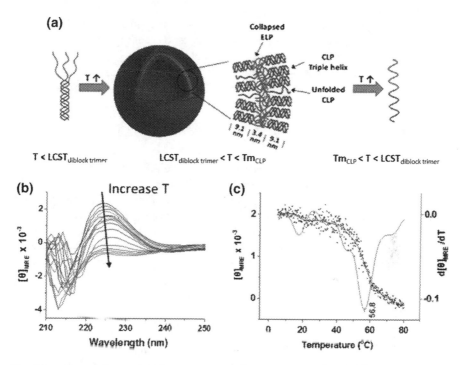

**Fig. 6.2** a Proposed assembly/disassembly and bilayer structure of ELP–CLP vesicles. b CD spectra showing representative full-wavelength scans for the ELP–CLP conjugate. c Thermal unfolding profile for the ELP–CLP conjugate; the first derivative of the unfolding curve with respect to temperature is shown in red

conjugates. However, the large size and polydispersity of the aggregates suggested that the hydrophobic interactions of the shorter (VPGFG)$_5$ were insufficient to form well-defined nanoparticles. In the end of this study, they indicated that changes of the CLP block stability, when balanced with the hydrophobicity of the ELP domain, could be employed to impart triggered assembly/disassembly under select conditions, promising in drug delivery, imaging, and materials modification.

Other intriguing biomedical applications such as in tissue engineering are also pervasive [11, 12]. Traditionally, covalently cross-linked hydrogels are mechanical favorable but have many side effects due to complex synthesis process, formation of toxic by-products, nonreversible cross-links. This irreversibility can potentially lead to cell death, induction of inflammatory responses, and ultimately failure of implanted devices in vivo. Though the non-covalently cross-linked hydrogels have many merits, the poor mechanical properties impeded their utility in biomedical applications. Ghoorchian and his coworkers innovatively developed a kind of micellar biohydrogels by genetically engineering an amphiphilic elastin-like polypeptides that contain a relatively large hydrophobic block and a hydrophilic terminus that can be cross-linked through metal ion coordination. To create the hydrogels, heat is firstly used to trigger the self-assembly of the polypeptides into monodisperse micelles that display transition metal coordination motifs on their corona, and subsequently cross-link the micelles by adding zinc ions. These hydrogels exhibit hierarchical structure, are stable over a large temperature range, and exhibit tunable stiffness, self-healing, and fatigue resistance.

## 6.2.2 Polysaccharides Regulated Assembly

Hydroxypropyl cellulose (HPC), a derivative of natural polysaccharide [13, 14] approved by the US FDA, has a lower critical solution temperature (LCST) in water ($\sim$ 41–45 °C) [15] and undergoes a phase transition in an aqueous environment from hydrophilic random-coil conformation to hydrophobic collapsed form as temperature increases above the LCST. The transition would cause dehydration of the polymer above the LCST due to both of increased hydrophobic interactions among polymer chains and decreased intermolecular hydrogen (H)-bonding interactions between ester groups in HPC and water molecules. Recently, a thermo-switchable polymeric photosensitizer (T-PPS) was designed by Park et al. [16]. to thermocontrol of the photoactivity of pheophorbide-a and synergistic combination with thermal cancer therapy. At temperatures below the polymer's LCST (43.4 ± 0.6 °C), the photosensitizer molecule on the HPC backbone aggregated in aqueous, resulting in quenching of the photoactivity of photosensitizer below LCST including body temperature region ($\sim$ 37 °C). However, as temperature increased to a hyperthermia cancer therapy temperature ($\sim$ 45 °C), the polymer dehydrated and formed fibrillary conformation. The dehydration created nonpolar microenvironment between the T-PPS molecules, which led to a monomeric state of the pheophorbide-a. As a result, the photoactivity of the photosensitizer was significantly enhanced in that condition, which was used for effective combinational cancer therapy with thermal cancer therapy.

Besides, chitosan that modified with hydroxybutyl groups also displays a LCST-type phase transition at temperatures 21–26 °C, depending on molar mass and degree of substitution. Chitosan is an attractive and easy obtaining biopolymer since it is produced by alkaline N-deacetylation of chitin that is found in the exoskeletons of shellfish and crustaceans. Hydroxybutyl chitosan (HBC) has shown to have excellent biocompatibility and minimal cytotoxicity [17]. HBC has been extensively utilized in medical field, such as post-operative treatment [17], tissue engineering [18], and therapeutics delivery [19]. The thermoresponsiveness of this chitosan derivative can be observed on the rheological properties of a 1 wt% solution that features a sol–gel transition temperature around 20 °C. These derivatives were used to construct 3D structures loaded with disk cells or mesenchymal stem cells that remained viable and proliferative for over 2 weeks without the addition of exogenous growth factors [20]. The sol–gel transition of the derivative was a fast process (within 60 s), implying the potential application for cell delivery to treatment of degenerative disk disease and potentially in tissue engineering.

## 6.2.3  Other Biopolymers Regulated Assembly

The more common situation with naturally occurring polymers is that they feature an upper critical solution temperature (UCST), above which the polymer is soluble and gelling [21] or form bundle [16] on cooling. For instance, collagen-like peptides (CLPs) [10, 22–25] or gum [26] adopt a random-coil conformation at high temperature, and this is actually the way to solubilize them, and, on cooling, a dehydrated triple-helix conformation forms via intermolecular hydrogen bonding, like gelatin does. The recent studies have illustrated that single-stranded CLPs have a strong propensity to bind native collagen via a strand invasion process. The high propensity of CLPs for collagen permits detection of minute quantities of collagen (e.g., 5 ng) with substantial promise for staining collagens in human tissues (e.g., skin, cornea, bone, and liver), especially those with high ECM turnover (e.g., prostate tumor xenografts, joints, and articular cartilage). Despite this widespread use, the utilization of CLPs as domains in responsive nanoparticles has been described in only a very limited number of reports. The most common applications of CLPs are as physical cross-linkers to form hydrogels or as gating materials for controlling drug release [27]. However, no application of UCST featuring polysaccharides in cell delivery has been reported, probably because their solubilization temperature is too high to be compatible with cell viability.

## 6.3  Synthetic Thermoresponsive Polymeric Nanoassemblies and Their Bioapplications

Despite highly biodegradable and biocompatible in nature, the batch-to-batch variability and broad molecular weight (MW) distributions of the natural polymers make them less attractive than synthetic polymers that are more reproducible and versatile. Since the earliest report of the thermal phase transition behavior of poly (N-isopropylacrylamide) (PNIPAAm) in 1967 by Scarpa et al., the "smart" thermoresponsive synthetic polymers have been extensively exploited. To date, several types of synthetic thermoresponsive polymers are developed. Some of the most prospecting synthetic polymers are discussed in this section.

## 6.3.1  Poly(N-Substituted (Meth)Acrylamide)S Regulated Assembly

Poly(N-substituted (meth)acrylamide)s are probably the most extensively studied thermoresponsive polymers [28]. In particular, poly(N-isopropylacrylamide) (PNIPAAm) with a sharp and easy tunable phase transition to near body temperature has garnered significant attention. Just like everything has minutes,

PNIPAAm possesses some inherent disadvantages, such as questionable biocompatibility, phase transition hysteresis, and a significant end group influence on thermal behavior. However, the toxicity only appears in oligomers or attributed to the residual monomer. The good thermosensitivity and independent of molecular weight and concentration [1] make it still irreplaceable and a prominent candidate in biomedical applications.

The phase transition temperature is determined by the balance between polymer–polymer interactions and polymer–water interactions. Structural factors that increase polymer–polymer attractive interactions decrease the transition temperature, while those that increase polymer–water attractive interactions increase the transition temperature. The transition temperature for PNIPAAm is then affected by the nature of the substitute groups [29]. Several copolymers of NIPAAm have been investigated to tune its transition temperature. The incorporation of more polar or ionic comonomers with NIPAAm causes the transition temperature of the resulting copolymer to shift to higher temperatures, due to the increased solubility from the polar groups. Inversely, the incorporation of less polar or hydrophobic monomers to PNIPAAm causes the transition temperature to shift to lower temperatures. For example, Hoffmann and Stayton groups modulated the phase transition of the thermoresponsive polymers by local polarity change in the irradiated azobenzene through the isomerization from trans-azobenzene to cis-azobenzene [30–32]. Besides, Katayama and coworkers reported that PNIPAAm-peptide containing the substrate peptide unit raised its transition temperature from 36.7 to 40°C in response to phosphorylation by activated protein kinase A (PKA) [33]. However, the same group, in another work, reported that the peptide sequence had a significant influence on phase transition temperature of the peptide–polymer conjugates [34]. For Peptide Ac-LRRASL-, which had an original net charge of +2 at physiological pH, the transition temperature of the conjugate decreased on phosphorylation (net charge became 0). In contrast, the transition temperatures of the conjugates with Peptide-ALRRASLE and Peptide Ac-DWDALRRASL-, both of which had neutral net charges, were greatly increased. This suggested that the transition temperature of the polymer was mainly governed by two factors: the change in polymer scaffold hydration and the phosphorylation resulted electrostatic repulsion among peptides.

It is noteworthy that copolymerization with monomers containing reactive functional groups such as carboxyl can hazard the sensitivity of the PNIPAAm copolymers [35]. The reason is that ionic moieties in the copolymers prevent intra- or intermolecular aggregation of the polymer chain, probably due to the increase of electrostatic repulsion or hydrophilicity [36]. Besides, the repeating isopropyl amide groups are very important to show very sensitive phase transition by rapid polymer aggregation over a narrow temperature range [37]. In a recent work, Aoyagi et al. maintained both the reactivity and temperature sensitivity of the polymer by an interesting molecular design [36]. In this work, a reactive monomer 2-carboxyisopropylacrylamide (CIPAAm) was copolymerized with NIPAAm. The important characteristic of this copolymer was that it was composed of the same polymer backbone and isopropylamide groups and some additional carboxyl groups. The phase transition of the copolymer occurred in a very narrow temperature range that was similar as PNIPAAm homopolymer. When both functional groups and thermosensitivity are needed, this molecular design concept is very useful.

The switchable property of thermoresponsive PNIPAAm can be used as a basic concept to manufacture attractive systems based on conformational/solubility changes in response to external temperature variations, resulting in thermoresponsive assembly/disassembly. The morphologies and sizes of these nanostructured assemblies can be tuned by controlling copolymer architecture (*e.g.*, linear, branched, graft), co-monomer sequence (e.g., random, gradient, block), composition, interactions between constituent monomers, and overall copolymer molecular weight. Similar to small-molecule surfactants, amphiphilic copolymers self-assemble into a variety of nanostructures in which the hydrophobic domains are shielded from the aqueous environment and the hydrophilic domains form a hydrated corona. The aggregated morphology is dictated primarily by the interfacial curvature within the aggregate and is a function of the packing parameter (p) described in equation $p = v/al$, where v is the volume of the hydrophobic chain, a is the optimal area of the hydrophilic head group, and l is the length of the hydrophobic chain. A lower packing parameter ($p \leq 1/3$) is indicative of high interfacial curvature and leads to spherical micelles, while worms and vesicles are observed as the packing parameter increases ($1/3 < p < 1/2$ and $p \leq 1/2$, respectively) (Fig. 6.3) [38]

Self-assembly of amphiphilic block copolymers can form a range of different nano-structures such as spherical micelles, vesicles, and cylinders. Among them, micellar and vesicle are the most designed structures. These structures form at a concentration larger than CMC when the temperature is higher than the phase transition temperature of the thermoresponsive polymer block. For example, Yang and his coworkers reported the use of Poly(ethylene oxide)-block-poly(N-isopropylacrylamide) to construct vesicle assemblies for drug delivery [39]. Besides, BAB triblock copolymers with a hydrophilic middle block flanked by two hydrophobic blocks have been designed to self-assemble into "flowerlike micelles." In a typical study, Hennink et al. [40] proved $pNIPAm_{16kDa}$-$PEG_{4kDa}$-$pNIPAm_{16kDa}$ can form "flowerlike micelles" by using static and dynamic light scattering measurements in combination with NMR relaxation experiments. However, these forms of assembly are labile under physiological environment. More stable structures are shell cross-linked micelles and core cross-linked micelles. For example, McCormick and coworkers have recently prepared hydrophilic–hydrophilic block copolymers, poly(N-(3-aminopropyl)methacrylamide hydro-chloride)-b-poly(N-isopropylacrylamide) (PAMPA-b-PNIPAM), and poly(2-(di-methylamino)ethylmethacrylate)-b-poly(N-isopropylacrylamide) PDMAEMA-b-PNIPAM) by reversible addition-fragmentation chain transfer (RAFT) polymerization [41]. These water-soluble block copolymers could directly form aggregates in water and that aggregate formation could be reversed by changing the solution temperature with respect to the LCST of PNIPAM (32 °C or higher, depending on the molecular weight of both blocks). The aggregates were trapped and stabilized by ionic cross-linking of the hydrophilic PAMPA block

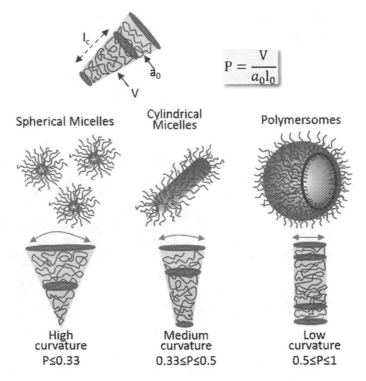

$$P = \frac{V}{a_0 l_0}$$

**Spherical Micelles**   **Cylindrical Micelles**   **Polymersomes**

High curvature
P≤0.33

Medium curvature
0.33≤P≤0.5

Low curvature
0.5≤P≤1

**Fig. 6.3** Various self-assembled structures formed by amphiphilic block copolymers in a block-selective solvent. The type of structure formed is due to the inherent curvature of the molecule, which can be estimated through calculation of its dimensionless packing parameter, p

through the addition of an oppositely charged polyelectrolyte that causes inter-polyelectrolyte complexation.

At a macroscopic level, so-called shape-changing materials, in which the material changes shape in the presence of stimulus and returns to its original shape upon removal of the stimulus, and shape-memory materials, which can be formed into a temporary shape but revert to an initial state upon the application of a specific stimulus, have been of great interest for a number of years [42]. The chief strategy used in designing these systems builds on a relatively simple model that relates amphiphilic balance to assembly structure. When the aggregation was driven by a non-covalent interaction, in thermoresponsive systems, aggregational rearrangement and micelle-vesicle transitions can be easily induced by temperature variation [43]. Thermally induced coil-globule transition of PNIPAM around its transition temperature (32 °C) has been utilized to adjust the amphiphilic balance of the block copolymers and the packing parameter p, leading to the morphology transition [44]. If the change in amphiphilic balance is drastic enough, this in turn leads to the destabilization of the originally stable structures and results in the reorganization and reassembly of the polymer aggregates into new thermodynamically favorable structures with parameters dictated by the new amphiphilic balance of the polymer chains [45]. Grubbs and his coworkers reported first Poly(ethylene oxide)-block-poly(N-isopropylacrylamide)-block-poly(isoprene)        (PEO-PNIPA-PI) copolymers that undergo a thermally induced reorganization in water from small spherical micelles ($\sim$ 10–20 nm diameter) to much larger spherical vesicles ($\sim$ 200 nm in diameter), though the transformation from micelles to vesicles takes approximately four weeks [46]. After cooling to 20 °C for about 48 h, the vesicles returned to assemblies with much smaller sizes. Then O'Reilly et al. demonstrated that diblock copolymer poly(tertbutylacrylate-b-N-isopropylacrylamide) realized the similar transition at 65 °C, which was faster but one week was needed [47]. Both of the results were successful in realizing the transition, but still suffered from the long duration for the transition and unsatisfactory reversibility. Replacing the hydrophobic poly(tertbutylacrylate) block by poly(methyl acrylate) with a lower transition temperature, O'Reilly's group achieved even faster morphology transition in about one day [48]. Recently, Grubbs et al. demonstrated that it was the strong interchain hydrogen bonding between the amide groups of PNIPAM that slowed down further rearrangement, resulting in a week-long transition [49]. This explanation is not in consistent with the well-known fact that the conformation change of PNIPAAm around its LCST (32 °C) takes place in seconds only [50]. Recently, Chen and Jiang realized the fast and reversible micelle-vesicle transition by greatly reducing the restriction to the mobility of the PNIPAAm chains imposed by the solid micellar core (Fig. 6.4) [51]. In this study, an asymmetrically end-modified PNIPAAm ($C_{12}$-PNIPAM-CA) with a short hydrocarbon chain ($C_{12}$) at one end and carboxyl group (CA) at the other was designed. Laser light scattering (LLS) measurement revealed that the transformation from micelles to vesicles can be realized within 30 min, while the reverse process only takes a few minutes. They proposed that when the temperature reaches LCST, the micelles as a whole serve as building blocks in constructing the vesicles via processes of combination, fusion, etc., in which only local conformation adjustment of PNIPAAm is involved.

The copolymer architecture is another important in determining the rate and nature of the transitions that occur in thermoresponsive block copolymer assemblies. The most widely investigated systems have a linear architecture in which the thermoresponsive block resides between a hydrophilic component and a hydrophobic component, changing the order and connectivity of these components changes the morphology of the assemblies. Besides, other architecture such as star type can result in different assemblies upon temperature variation. It is worth to known that developing a better understanding of the effects of both polymer composition and structure on structural evolution in these systems will allow optimization of response time for specific applications [42].

Except constructing delivery systems, thermoresponsive polymers can also be utilized to fabricate artificial protein-mimicking assemblies. Nature is powerful at building nanostructures with great complexities via precisely controlling multi-stage self-assembly processes. This highly specific, hierarchical and cooperative chemistry enables the living systems to maintain themselves in a dynamic equilibrium. In molecular biology, molecular chaperones are proteins that assist the

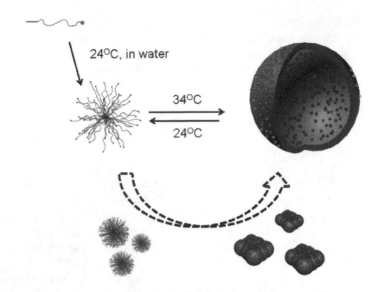

**Fig. 6.4** Schematic illustration of the thermally induced morphology transition

non-covalent folding/unfolding or assembly/disassembly of other proteins. Heat-shock proteins (Hsps), the most common kind of molecular chaperones, play an effective role in promoting proper protein conformation, as well as inhibiting unwanted protein aggregation, in response to elevated temperature or other stimulus. Inspired by the native occurring chaperones, especially the GroEL/GroES system, Shi and his coworkers developed a series of artificial chaperones based on mixed-shell polymer micelles (MSPM) [52]. A microphase-separated surface formed by the reversible PNIPAAm coil-globule transition through switching temperature, preventing unwanted aggregation of model protein carbonic anhydrase B and assisting the refolding of denatured proteins (93% in efficiency). The same group, in another work, demonstrated that the use of the artificial chaperone to maintain Aβ homeostasis by a combination of inhibiting Aβ fibrillation and facilitating Aβ aggregate clearance and simultaneously reducing Aβ-mediated neurotoxicity, serving as a suppressor of Alzheimer's disease (Fig. 6.5) [53]. In this work, the artificial chaperone was constructed through the co-assembly of block copolymers poly(lactide-b-poly(ethyleneoxide) (PLA-b-PEG) and poly-(lactide)-b-poly(N-isopropylacryamide) (PLA-b-PNIPAAm), in which the mixed-shell polymer micelle shared PLA core and a homogeneously mixed PEG and PNIPAAm shell. Upon raising the temperature up to the normal body temperature, the PNIPAAm chains in the micellar shell spontaneously underwent a hydrophilic to hydrophobic transition and collapse, forming hydrophobic domains on the PCL core. These hydrophobic domains acted as anchors for interacting with hydrophobic Aβ monomers or its intermediate oligomeric aggregates, while the hydrophilic and stretched PEG chains could create a protective barrier layer to prevent excessive amounts of Aβ adsorption and the aggregation. Further, the area of the PNIPAAm

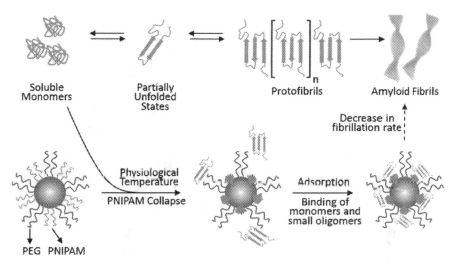

**Fig. 6.5** Representation of the possible mechanism of interaction between MSPMs and Aβ peptide on the kinetics of amyloid fibrillation process

domain can be precisely tuned by variation of the ratio of the two copolymers, thus modulating the surface properties of the mixed-shell polymer micelles and control the effect of artificial chaperone on the Aβ fibrillation progress. More interestingly, the microphase separated surface have the advantageous property of long blood circulation [53, 54].

Thermosensitive PNIPAAm can also be used to construct thermometers to measure the thermal fluctuations in cellular processes. Since 2003, Uchiyama and coworkers have developed a variety of fluorescent polymeric thermometers by taking advantage of the effect that the quantum yields of polarity-sensitive fluorescent dyes can be dramatically enhanced within hydrophobic microenviron ments when polymer chains are subjected to thermo-induced aggregate formation [55]. In a typical example, they copolymerized NIPAAm with water-sensitive fluorescent monomers, yielding a new highly hydrophilic PNIPAAm nanogel material that exhibited temperature-dependent fluorescence [56]. Intracellular thermometry tests using nanogel-implanted monkey kidney stem cells (COS7) showed that the intracellular temperature variations associated with biological processes can be monitored with a temperature resolution of better than 0.5 °C in the range between 27 and 33 °C, which was independent of pH and the protein environment. However, this nanogel-based thermometer could measure only the average temperature of the whole cell, because the relatively large size (62 nm or more in hydrodynamic diameter) and low hydrophilicity (aggregation was induced at temperatures higher than 27 °C) of nanogel-thermometer hindered the dispersion of this temperature probe throughout the cell. Recently, they developed a linear fluorescent polymeric thermometer that could diffuse throughout the cell and applied it to intracellular temperature mapping, where the fluorescence lifetime of

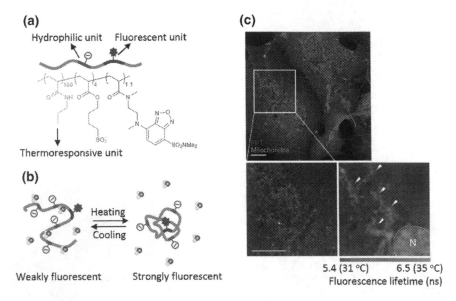

**Fig. 6.6** Intracellular temperature mapping with a fluorescent polymeric thermometer. **a** Chemical structure of the fluorescent polymeric thermometer. **b**, Functional diagram I an aqueous medium. **c**, Confocal fluorescence images of FPT (green) and Mito Tracker Deep Red FM (Red upper and left lower) and fluorescence lifetime image of FPT (right lower). *Arrows*, local heat production. *N*, nucleus

this thermometer was adopted as a temperature-dependent variable (Fig. 6.6) [57]. This polymers consisted a thermoresponsive unit, a fluorescent unit, and an anionic hydrophilic unit to prevent the precipitation, and the temperature of difference cell organelles is successfully measured.

## 6.3.2 Poly(Alkyloxide)S Regulated Assembly

Poly(alkyloxide)s are another type of extensively studied thermoresponsive polymers because of their neutral, nontoxic, and nonimmunogenic character. Poly (ethylene oxide) (PEO) [or Poly(ethylene glycol) (PEG)] and poly(propylene oxide) (PPO) and their copolymers are well known for their temperature-responsive behavior. Amphiphilic balance in structures is the leading driving force for this temperature-responsive feature. PEG–b–PPO copolymers are of especial interest for their reverse thermal gelation behavior, which arises from the effect of the LCST-mediated transition on solution viscosity. Solutions of these polymers in water exhibit a dramatic viscosity increase with temperature, forming semisolid gels when heated above the LCST. A large variety of PEG and PPO block copolymers known as Pluronics, Poloxamers, and Tetronics are commercially available. Aqueous solutions of PEG–b–PPO–b–PEG (pluronics or poloxamers) demonstrated phase transitions from sol to gel at 5–30 °C and gel to sol at 35–50 °C with the temperature increasing monotonically over the polymer concentration range of 20–30 wt% [58, 59]. As temperature increases, equilibrium shifted from unimers to spherical micelles, leading to an increase in micelle volume fraction ($\varphi$). When $\varphi$ was larger than 0.53, the system becomes a gel by micelle packing [60]. The transition mechanism from gel-to-sol is related to the shrinkage of PEG corona of the micelles because of temperature effects on PEG solubility and interaction of PEG chains with the PPO hard core.

The sol-to-gel transition makes aqueous PEG–b–PPO–b–PEG an attractive solution in drug, cells, or proteins delivery applications as an in situ gel-forming device. The minimally invasive, in situ gelling injection system creates an advantageous alternative to surgical procedure. However, the physical cross-linking between micelles is so weak that the gel integrity of PEG–PPO–PEG does not last more than several hours. To improve gel durability and biocompatibility, poly (ethylene glycol–b–(DL-lactic acid–co–glycolic acid)–b–ethylene glycol) (PEG–b–PLGA–b–PEG) triblock copolymers as a novel injectable depot system was designed. PEG–b–PLGA–b–PEG is free-flowing sol at room temperature and becomes a gel at body temperature. The sol-to-gel transition temperature of PEG–b–PLGA–b–PEG triblock copolymers in aqueous solution could be controlled over a temperature range of 15–45 °C by changing the molecular parameters of PEG–b–PLGA–b–PEG triblock copolymers, such as PLGA length, PEG length and ratio of DLLA to GA in the middle block. The more hydrophobic the polymer, the stronger the shear stress required to make the gel flow. In situ gel formation was confirmed by subcutaneous injection of PEG–b–PLGA–b–PEG triblock copolymer aqueous solutions (33 wt%) into rats, and results showed that its integrity persisted for more than one month. Owing to the triblock topology, the PEG–b–PLGA–b–PEG polymers have limitations in terms of molecular weight and degradation profile to show the sol-to-gel transition in an undesired range of ∼ 10–30 °C. To solve this issue, graft copolymers of PEG–g–PLGA and PLGA–g–PEG with sol-to-gel transitions at ∼ 30 °C were developed. PEG–g–PLGA copolymers have hydrophilic backbones and form gels with short durability, whereas PLGA–g–PEG copolymers have hydrophobic backbones and form much more durable gels. By mixing the two copolymers, the durability of a gel can be controlled from one week to three months.

Normally, PEG and PPO copolymers have consisted of linear high molecular weight polymers or block copolymers. Recently, there has been a great deal of interest in (co)polymers with short oligo(ethylene glycol) (OEG) side chains. The transition temperature of this kind of thermoresponsive polymer can be tuned by the size of the OEG segment and also by copolymerization with 2-(2-methoxyethoxy) ethyl methacrylate (MEO$_2$MA). OEG-based polymers with intermediate side chain lengths generally exhibit cloud points in aqueous solution that are lower than high molecular weight PEG due to the increase hydrophobicity of the hydrocarbon polymer backbone. For instance, poly(2-(2-methoxyethoxy)ethylmethacrylate) (PMEO$_2$MA) and PMEO$_3$MA have transition temperatures around 26 and 52 °C, respectively. Poly(oligo(ethylene glycol) methacrylate)s (POEGMA) with longer side chains (4–9 ethylene oxide units) exhibit transitions in the range 60–90 °C. A variety of thermoresponsive OEGMA-based polymers and their use in diverse biologically relevant applications have recently been highlighted by Lutz [61]. The thermal phase transition temperature of the copolymers can be fine-tuned by varying ratios of MEO$_2$MA and OEGMA. This precise control of transition temperature allows for the synthesis of P(MEO$_2$MA-co-OEGMA) brushes as described in the thermoresponsive hybrid part.

There is a drawback of OEGMA-based polymers; that is, they display a salting-out effect. Alexander and coworkers have demonstrated that the salt type and concentration in the polymer solution had a major impact on supramolecular organization. In non-saline-based buffers, this salting-out effect was not exhibited, and the corresponding transition temperatures were similar to those observed in aqueous solution. However, in the cell culture media, which contained a mixture of salts along with various proteins, amino acids, and antibiotics, the transition temperatures would decrease and hazard biological applications.

### 6.3.3 Polyphosphoester-Based Thermoresponsive Polymers Regulated Assembly

Polyphosphoesters (PPE)-based thermoresponsive polymers are biodegradable materials that emerge as a promising class of novel biomaterials [62, 63]. Polyphosphoesters-based thermoresponsive polymers are polymerized via organocatalyzed ring-opening polymerization of cyclic phosphoester monomers. The transition temperatures of the copolymers can be tuned by varying the copolymer composition. PPEs are able to undergo hydrolytic degradation by acid/ base or enzymatic catalyzed routes, in which the degradation rate of can be regulated by controlling the chemical structure of the backbone and pendant groups. Thermoresponsive and biodegradable polymer nanoparticles were prepared and utilized in a myriad of biomedical applications such as drug, gene, and protein carriers.

### 6.3.4 Other Thermoresponsive Synthetic Polymers Regulated Assembly

A wide variety of other classes of polymers with LCST-type behavior has been reported. For example, water-soluble poly(N-vinylalkylamide)-based polymers show dramatic thermoresponsive properties. Akashi et al. synthesized a series of poly(N-vinylalkylamide) derivatives having different alkyl side chain lengths. These polymers have amide side chains with the nitrogen bonded directly to the polymer backbone. Poly(N-vinylisobutyramide) (PNVIBA), a structural isomer PNIPAAm, exhibits a sharp thermal transition at 39 °C, and the cloud point is strongly dependent on the polymer and salt concentration in solution. Although it can be argued the hydrophilic–hydrophobic balance is the same as PNIPAAm, the transition temperature of PNVIBA is higher, which may be attributed to the microstructure of the hydrated polymer.

Poly(methyl vinyl ether) (PMVE) has a transition temperature exactly at 37 °C, which makes it very interesting for biomedical application. The polymer exhibits a typical type III demixing behavior, which is in contrast to the thermal behavior of PNIPAAm.

Poly(N-ethyl oxazoline)s have a transition temperature around 62 °C, which is too high for any drug delivery application. However, we recently prepared a double thermoresponsive system by graft polymerization of EtOx onto a modified PNIPAAm backbone. Currently, these systems are explored for their potential in drug delivery, because they tend to form micelle aggregates above the LCST. Unfortunately, the poly(oxazoline) chemistry has the disadvantage that it is not very tolerant against unprotected functionalities.

## 6.4 Self-assembled Thermoresponsive Hybrid Materials for Tumor Imaging and Therapy

Some of the smart nanosystems presented here are not only composed of thermoresponsive units, but also of other building blocks that have a structural function and contribute to the smartness of the system acting as triggers or labels. By far, a variety of smart hybrid materials have been established and applied in a wide range of areas including bioseparations, diagnostics and biosensing, drug/siRNA delivery, enzymatic activity modulation, cell culture processes, and CTC capturing.

### 6.4.1 Biohybrid Copolymers Regulated Assembly

In 1970s, Ringsdorf proposed the concept of polymer–biomacromolecule conjugates to narrow the gap between polymer chemistry and biology. The benefits of attaching synthetic polymers to proteins are multifold, as it renders the biological component with responsive nature, which may allow fine tuning of the solubility, stability, and/or even bioactivity of the biomacromolecules under control of environmental temperature variation.

Increasing focus on protein therapeutics has led to a rapid growth in research aimed at preparing polymer–protein conjugates. Hoffman and his coworkers carried out pioneering work in the early development of thermoresponsive polymer–protein conjugates in 1980s. They first reported bioseparation of certain proteins from blood serum using a "monomer conjugation plus copolymerization" strategy. After then, they further developed the thermally induced phase separation immunoassay by conjugating poly(N-isopropylacrylamide) (PNIPAAm) to monoclonal antibody (mAb). Thermosensitive PNIPAAm underwent a sharp reversible coil-globule phase transition in water at temperature above/below its LCST ($\sim 32$ °C), being hydrophobic above this temperature and hydrophilic below it. Thus, the PNIPAAm-mAb conjugates enabled the partition and separation of target antigens in water on the basis of solubility. Since then, a myriad of works are performed to conjugate thermoresponsive polymers to proteins [64, 65], peptides [37], DNAs [66], and polysaccharides [67] for the bioapplications in drug/protein delivery, bioseparation, enzyme activity manipulation, control of protein binding, diagnostics, and tissue engineering.

Except for bioseparation and controlling the activity of bioactive biological components upon the aggregation and disaggregation of the biohybrid materials, constructing drug/protein delivery systems is another important application. Recently, Thordarson group demonstrated the use of thermoresponsive fluorescent protein-polymer bioconjugates to prepare polymersomes for capturing of and reporting on drug and protein payloads (Fig. 6.7) [68]. A free cysteine residue at position 119 was designed on the protein, and then bioconjugated to the maleimide-terminated PNIPAAm through a grafting to method. The PNIPAAm underwent hydrophilic to hydrophobic phase transition upon heating above 36.5 °C (its LCST), which triggered the aggregation of the biohybrid and formed polymersomes. The as-formed polymersomes spatial controlled the encapsulation of DOX and light-harvesting PE545 protein. Besides, the fluorescence property of the protein allowed a combination of fluorescence lifetime imaging microscopy (FLIM) and Forster resonance energy transfer (FRET; FLIM-FRET) which precisely reveals the location of the encapsulate within the polymersomes.

Biological clues are significant importance in regulating a variety of important processes in living objectives, such as enzymes regulating ligation and degradation of their substrates, glutathione (GSH)-maintaining cellular reduction environment. Using biological triggers to control the LCST-switch and subsequent aggregation is very promising in biomedical applications, such as detecting enzyme activity and monitoring disease progression. Substrate peptide of protein kinase A (PKA) with different structures was conjugated with PNIPAAm by Katayama et al. [34]. PKA is a cAMP-dependent protein kinase, and which sustained activation is related to various diseases such as melanoma and colon cancer. The conjugates containing PNIPAAm and the PKA substrate peptide motifs showed an increase in their LCST from 38 °C to 46.5 °C in response to phosphorylation by activated PKA. At temperatures above their LCSTs, the conjugates formed micellar structures. The poly(ethylene glycol) was used as the outer shell and PNIPAAm with N-methacryloyl bearing substrate peptide sequence was designed as inner core. An LCST switch of the inner core moiety under the control of PKA signal suggested that encapsulation of drugs into this micelle might result in a smart drug delivery system that communicated with an intracellular signal. Recently, our group reported an endogenous stimuli-induced-aggregation (eSIA) approach to construct

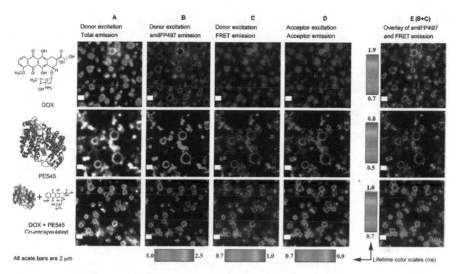

**Fig. 6.7** Fluorescence lifetime imaging microscopy (FLIM) images of PNIPAM-b-amilFP497 polymersomes formed at 408 C with encapsulated doxorubicin (DOX, top row), phycoerythrin 545 (PE545, middle row), and a co-encapsulated mixture of DOX and PE545 (bottom row). Images in columns A, B, C, and the overlay in E (B + C) were obtained using 405 nm excitation, whereas images in columnD were obtained using 532 nm excitation. The relevant lifetime color scales (in nanoseconds) are shown under columns B–D, while for the images in column E, the color scale is shown to the left of these images. **a** Grayscale fluorescence image of 485/32 nm (32 nm bandpass filter centered at 485 nm) and 617/73 nm (73 nm bandpass filter centered at 617 nm) channels overlaid (total emission). **b** 485/32 nm channel corresponding to amilFP497 emission with a lifetime colormap from 0 to 2.5 ns. **c** 617/73 nm channel corresponding to the FRET-based acceptor emission with a colormap from 0.7 (similar to PE545) to 1.0 ns (similar to DOX). **d** 617/73 nm channel corresponding to acceptor emission using 532 nm excitation with a colormap from 0.7 (similar to PE545) to 0.9 ns (similar to DOX). E) Overlay of B + C; the amilFP497 emission in the 485/32 nm channel and the FRET acceptor emission in the 617/73 nm channel (false red). Detailed lifetime analysis (Tables S2–S7) and phasor analysis of this data is in the supporting information. The scale bar represents 2 mm

functional nanoaggregates for in situ sensing and monitoring cellular physiological process (Fig. 6.8) [69]. In this research, we designed a series of thermosensitive polymer–peptide conjugates (PPCs), which were composed of three moieties, i.e., a thermoresponsive polymer backbone, a grafted responsive peptide and a signal-molecule label. The PPCs were initially polymer coil at physiological temperature, whereas formed fluorescent nanoaggregates under a specific intracellular stimulus. The as-formed fluorescent nanoaggregates exhibited long-term retention effect in cells, which enabled specific-identification and quantification of endogenous factors. This general approach is promising for high-performance in situ sensing and dynamic monitoring of disease progression in living subjects.

Another fascinating application of the thermoresponsive biohybrid is to fabricate artificial antibodies [70] or vaccines [71]. Antibody-targeted polymeric nanoparticles (immunoparticles, approximately 320 nm in diameter) were prepared by

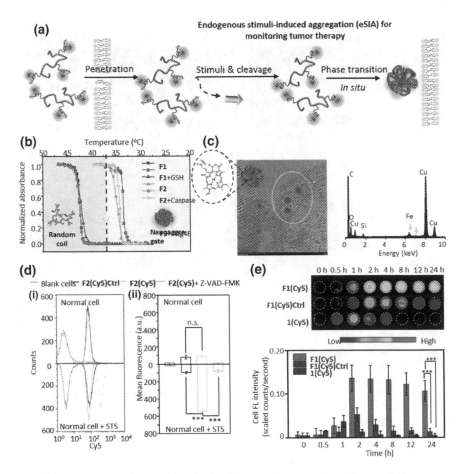

**Fig. 6.8** **a** Schematic representation of stimuli-instructed construction of controllable nanoaggregates for monitoring tumor therapy response. **b**, LCST profiles before and after the treatment of stimuli. **c** Visualization and identification of nanoaggregates formation in cells. Enhanced uptake **d** and retention **e** after nanoaggregates formation

Goodall et al. [70]. through self-assembly driven of the thermoresponsive polymer that covalently attached with an EGFR-targeting scFv antibody fragment. The binding property of the scFv to recombinant EGFR and to native EGFR on MDAMB468 cells was determined using surface plasmon resonance (SPR) and flow cytometry, respectively. The thermoresponsive polymer did not compromise the binding affinity of the scFv to EGFR. Additionally, the immunoparticle was further stabilized by cross-linking the scFv with glutaraldehyde, which did not impact the binding affinity to EGFR. The thermally controlled self-assembly of immunoparticles from antibody-conjugated polymers provided a novel strategy to fabricate targeted nanoparticles. This methodology further provides a rapid and versatile way to construct immunoparticles with theranostic capability by

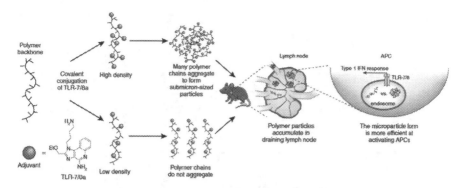

**Fig. 6.9** Polymeric adjuvants are synthesized by conjugating a small-molecule TLR-7/8 agonist (TLR-7/8a) to a hydrophilic polymer backbone. At higher conjugation densities, the hydrophobicity of the agonist causes self-aggregation of the polymers and the formation of submicron-sized particles. These particles, which better traffic to the lymph node and more effectively activate APCs in comparison to non-aggregated polymer chains, stimulate a type 1 interferon (IFN) response and promote a robust adaptive immune response to model antigens

combining different antibodies with different thermosensitive polymers. In another case, Lynn and coworkers attached protein antigens to PNIPAAm scaffolds and co-assembled with polymer–TLR-7/8a to form mate materials with different physicochemical properties [71]. They systematically evaluated how different physicochemical properties of the hybrid materials influenced the location, magnitude, and duration of innate immune activation in vivo (Fig. 6.9). Experimental results showed that particle formation was the most important factor for restricting adjuvant distribution and prolonging activity in draining lymph nodes. Besides, the particulate structures improved the pharmacokinetic profile by reducing morbidity and enhancing vaccine immunogenicity that induced antibodies and T cell immunity.

Harnessing responsive nature of the thermosensitive polymers to regulate membrane receptor oligomerization is a new direction for their applications [37]. Receptor clustering on cell membranes controls lots of cellular activity, such as cell apoptosis, proliferation, and tumorigenesis. We recently reported a thermocontrolled strategy for realizing in situ conformational transition of polymer–peptide conjugates at cell surfaces to manipulate and monitor HER2 receptor clustering, which finally resulted in effective breast cancer cell proliferation inhibition (Fig. 6.10). In this study, functional paring motifs (bis(pyrene) conjugated HER2-targeting peptides) were covalently linked to a PNIPAAm polymer to incorporate temperature control manner to HER2 targeting peptide. At 40 °C, a temperature above the LCST, PNIPAAm polymers drive the assembly and act as a "shield" to block the aggregation of HBP. Upon cooling down to 35 °C, polymers hydrated, extended, and exposed HBP to water, then the bis(pyrene)s aggregated and dragged HER2 receptors to oligomerize. Bis(pyrene) exhibited aggregation-induced fluorescence (AIE) feature, allowing in situ monitoring

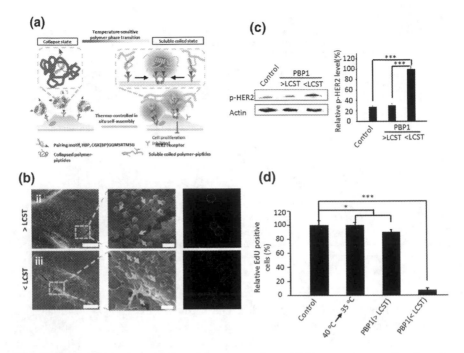

**Fig. 6.10** **a** Schematic representation of thermocontrol over HBP aggregates on cell surfaces for manipulating and monitoring HER2 receptor clustering. **b** SEM and confocal images to visualize PBP1 structural transition on SK-BR-3 cell surfaces. Yellow arrows, irregular protrusions; red arrows, collapsed state of PBP1 at 40 °C (> LCST). Scale bar: 5 μm (left lane) and 1 μm (right lane). **c** Western blot analysis of HER2 phosphorylation induced by HBP aggregation at 40 °C (> LCST) and 35 °C (< LCST). **d** Histogram shows significant decrease of EdU-labeled cells in the PBP1 treated SK-BR-3 cells at 35 °C (<LCST) and temperature change alone had no effect on cells

receptor clustering. Ultimately, HER2 receptor clustering leads to cytoplasmic domain phosphorylation, which further result in high-performance inhibition of cancer cell proliferation. Noticeably, this is the first example of thermocontrolling assembly/disassembly of cell surface receptors by using thermosensitive polymer-peptides under physiological conditions to regulate biological processes. We conjecture that this convenient approach has the potential to be applied for molecule-targeted tumor therapy.

### 6.4.2 Organic/Inorganic Hybrid Materials Regulated Assembly

Organic/inorganic hybrid materials can serve and surpass their parent materials, making them hold a myriad of attractive bioapplications, such as biosensing, bioimaging, photothermal therapy (PTT), and drug delivery. Gold nanoparticles

(AuNPs), gold nanorods (AuRds), and many other inorganic nanoparticles exhibit strong size-dependent absorbance and scattering in the visible to near-infrared region, which is caused by localized surface plasmon resonance (SPR). Generally, the plasmonic peaks of 10–100 nm-sized AuNPs are approximately 520–540 nm, which fall outside the tissue transparency window (650–1000 nm) and hinder their biomedical imaging and therapy applications. Toward this issue, several strategies are developed to regulate the assembly of AuNPs, which can induce the near field coupling of surface plasmons between adjacent particles and result in redshift of the plasmonic peaks in a variety of degrees. Of particular interest is the reversible assembly of the plasmonic particles, which is expected to enable dynamic tuning of the surface plasmon coupling in response to the external stimuli, and therefore produce active optical materials. The reversible property of the thermoresponsive polymers in regulating assembly makes them the most suitable materials to do this work.

The optical signals of assembled AuNPs/AuRds are known to be influenced by the dimensions of the assembled structures, distances, and cluster sizes between the nanoparticles. Besides, there are three forces, i.e., van der Waals attraction, electrostatic repulsion, and short-range repulsion in controlling AuNPs assembly in the thermosensitive polymer-AuNPs systems. Considering this, Liu et al. [72]. succeeded in thermocontrolling the assembly and disassembly of bis (p-sulfonatophenyl)-phenylphosphine (BSPP)-coated AuNPs through manipulation of electrostatic interactions of the particles by temperature variation, which also further dynamically and reversibly tuned the optical properties of the AuNPs. Besides, Meyerbroker and coworkers reported temperature-dependent changes in optical signals of AuNPs that adsorbed on a ultrathin poly(ethylene glycols) hydrogel film by using its swelling/deswelling properties [73]. Meanwhile, several other studies have attempted to acquire thermoreversible changes in the optical signals of AuNPs by using thermoresponsive materials. However, the optical bandwidths that were regulated by temperature variation have been too narrow (mostly 30–50 nm), lowering the detection sensitivities of this systems. In a recent work, reported by Lim et al. [74] spherical AuNPs (15–51 nm in diameter, displaying an extinction peak wavelengths of 520–530 nm) were assembled on thermoresponsive nanogels. The assembly structures of the AuNPs in the hybrid colloids were reversibly manipulated from temperature-dependent swelling/ shrinking behaviors of the nanogels. Large AuNPs that introduced on the nanogel resulted in the largest red shift of the optical signal to near-infrared region ($\sim 900$ nm). Also, in another work, Ding et al. [50] showed that 60 nm AuNPs had a largest 200 nm redshift in extinction peaks, while the 100 nm particles displayed a slightly blueshift. This was associated with increased electrostatic repulsion. They further investigated the effect of PNIPAAm concentration, heating time on the final cluster sizes. Results showed that a middle concentration of 20 µM PNIPAAm resulted in the largest peak redshift; while, excess PNIPAAm can increase the coating thickness, which spaced the AuNPs cores further apart and decreased the maximum redshift. The plasmon resonance peak redshifted with heating time last, yet small blueshifted at the longest times due to the rearrangement of AuNP clusters from nonsphererial aggregates into more compact arrangements.

For biological applications, the phase transition temperatures of the thermosensitive hybrid materials should be fine-tuned. Normally, the transition temperature of the hybrids is determined by the curvature of the inorganic particle, polymer composition, the polymer coating density, and the polymer chain length. Gibson group reported that small gold nanoparticles (<50 nm) that bearing a thermosensitive polymer corona display a strong size dependence on their transition temperatures. Larger particle had lower phase transition temperature. The underlying mechanism of this cloud point depression is probably based on the particle curvature differences, which influences the polymer packing density. As nanoparticle size increases, the curvature decreases leading to an increase in the crowding of polymer chains, assuming equal surface coverage. This crowding will increase the polymer–polymer interactions and thus decrease the LCST in a similar manner to increasing the solution concentration [75]. The same group, in another work, [76] demonstrated the thermoresponsive properties of the binary mixtures of PPEGMA-coated gold nanoparticles could be tuned by varying the mass fraction of the constituent particles. Based on this discovery, they further explored the feasibility of the cooperative temperature-induced aggregation behavior of suitable pairs of thermoresponsive polymers to direct the self-assembly of gold nanoparticles onto solid substrates. Besides, they also reported an extensive investigation into the use of blending (i.e., mixing) as a powerful tool to modulate the transition temperature of poly(N-isopropylacrylamide) (PNIPAM)-coated gold nanoparticles [77]. By simply mixing two nanoparticles of different compositions, precise control over the transition temperature can be imposed. Cooperative aggregation of differently sized nanoparticles (with different cloud points) was achieved.

## 6.5   Concluding Remarks and Outlook

This chapter focused on the thermoresponsive material-regulated assemblies from the viewpoints of their molecular design, self-assembly mechanism, formed structures, and biofunctions. We reviewed a variety of thermoresponsive polymeric assemblies, including naturally occurring thermoresponsive polymeric assemblies, synthetic thermoresponsive assemblies, and hybrid-type thermoresponsive assemblies. Their biomedical applications were also demonstrated. Fabricating thermoresponsive assemblies in vitro for in vivo applications was the main trend in this field. However, in vivo assembly using thermoresponsive materials had sprouted in our group recently. Constructing self-assemblies in vivo can better mimic biological processes, compete or even surpass nature's performance. The development of "in vivo assembly" thermoresponsive materials needs considering the polymers' phase transition temperature and sensitive temperature responsiveness. All these parameters are encoded in the molecular level.

**Acknowledgements** This work was supported by the National Natural Science Foundation of China (51573032) and Science Fund for Creative Research Groups of the National Natural Science Foundation of China (11621505), Key Project of Chinese Academy of Sciences in Cooperation with Foreign Enterprises (GJHZ1541), CAS Key Research Program for Frontier Sciences (QYZDJ-SSW-SLH022), and CAS Interdisciplinary Innovation Team.

# References

1. Scarpa JS, Mueller DD, Klotz IM (1967) Slow hydrogen-deuterium exchange in a non-alpha-helical polyamide. J Am Chem Soc 89(24):6024–6030
2. Ward MA, Georgiou TK (2011) Thermoresponsive polymers for biomedical applications. Polymers 3(4):1215–1242
3. Fukumura D, Jain RK (2007) Tumor microenvironment abnormalities: causes, consequences, and strategies to normalize. J Cell Bio Chem 101(4):937–949
4. Foster JA, Bruenger E, Gray WR, Sandberg LB (1973) Isolation and amino acid sequences of tropoelastin peptides. J Biol Chem 248(8):2876–2879
5. Martin L, Castro E, Ribeiro A, Alonso M, Rodriguez-Cabello JC (2012) Temperature-triggered self-assembly of elastin-like block co-recombinamers: the controlled formation of micelles and vesicles in an aqueous medium. Biomacromol 13(2):293–298
6. Li NK, Quiroz FG, Hall CK, Chilkoti A, Yingling YG (2014) Molecular description of the lcst behavior of an elastin-like polypeptide. Biomacromol 15(10):3522–3530
7. Urry DW (1997) Physical chemistry of biological free energy transduction as demonstrated by elastic protein-based polymers. J Phys Chem B 101(51):11007–11028
8. Quiroz FG, Chilkoti A (2015) Sequence heuristics to encode phase behaviour in intrinsically disordered protein polymers. Nat Mater 14(11):1164–1171
9. MacKay JA, Chen M, McDaniel JR, Liu W, Simnick AJ, Chilkoti A (2009) Self-assembling chimeric polypeptide doxorubicin conjugate nanoparticles that abolish tumours after a single injection. Nat Mater 8(12):993–999
10. Luo T, Kiick KL (2015) Noncovalent modulation of the inverse temperature transition and self-assembly of elastin-b-collagen-like peptide bioconjugates. J Am Chem Soc 137(49):15362–15365
11. Ghoorchian A, Simon JR, Bharti B, Han W, Zhao X, Chilkoti A, López GP (2015) Bioinspired reversibly cross-linked hydrogels comprising polypeptide micelles exhibit enhanced mechanical properties. Adv Funct Mater 25(21):3122–3130
12. Inostroza-Brito KE, Collin E, Siton-Mendelson O, Smith KH, Monge-Marcet A, Ferreira DS, Rodríguez RP, Alonso M, Rodríguez-Cabello JC, Reis RL, Sagués F, Botto L, Bitton R, Azevedo HS, Mata A (2015) Co-assembly, spatiotemporal control and morphogenesis of a hybrid protein–peptide system. Nat Chem 7(11):897–904
13. Kita R, Kaku T, Kubota K, Dobashi T (1999) Pinning of phase separation of aqueous solution of hydroxypropylmethylcellulose by gelation. Phys Lett A 259(3–4):302–307
14. Porsch C, Hansson S, Nordgren N, Malmström E (2011) Thermo-responsive cellulose-based architectures: tailoring LCST using poly(ethylene glycol) methacrylates. Polym Chem 2(5):1114
15. Tan J, Kang H, Liu R, Wang D, Jin X, Li Q, Huang Y (2011) Dual-stimuli sensitive nanogels fabricated by self-association of thiolated hydroxypropyl cellulose. Polym Chem 2(3):672–678
16. Park W, Park SJ, Cho S, Shin H, Jung YS, Lee B, Na K, Kim DH (2016) Intermolecular structural change for thermoswitchable polymeric photosensitizer. J Am Chem Soc 138(34):10734–10737

17. Wei C-Z, Hou C-L, Gu Q-S, Jiang L-X, Zhu B, Sheng A-L (2009) A thermosensitive chitosan-based hydrogel barrier for post-operative adhesions' prevention. Biomaterials 30 (29):5534–5540

18. Zhang K, Qian Y, Wang H, Fan L, Huang C, Yin A, Mo X (2010) Genipin-crosslinked silk fibroin/hydroxybutyl chitosan nanofibrous scaffolds for tissue-engineering application. J Biomed Mater Res A 95(3):870–881

19. Dang JM, Sun DDN, Shin-Ya Y, Sieber AN, Kostuik JP, Leong KW (2006) Temperature-responsive hydroxybutyl chitosan for the culture of mesenchymal stem cells and intervertebral disk cells. Biomaterials 27(3):406–418

20. Dang JM, Sun DD, Shin-Ya Y, Sieber AN, Kostuik JP, Leong KW (2006) Temperature-responsive hydroxybutyl chitosan for the culture of mesenchymal stem cells and intervertebral disk cells. Biomaterials 27(3):406–418

21. Delair T (2012) In situ forming polysaccharide-based 3D-hydrogels for cell delivery in regenerative medicine. Carbohyd Polym 87(2):1013–1019

22. Krishna OD, Kiick KL (2010) Protein- and peptide-modified synthetic polymeric biomaterials. Biopolymers 94(1):32–48

23. Luo T, He L, Theato P, Kiick KL (2015) Thermoresponsive self-assembly of nanostructures from a collagen-like peptide-containing diblock copolymer. Macromol Biosci 15(1):111–123

24. Li L, Luo T, Kiick KL (2015) Temperature-triggered phase separation of a hydrophilic resilin-like polypeptide. Macromol Rapid Commun 36(1):90–95

25. Krishna OD, Wiss KT, Luo T, Pochan DJ, Theato P, Kiick KL (2012) Morphological transformations in a dually thermoresponsive coil-rod-coil bioconjugate. Soft Mat 8 (14):3832–3840

26. Coutinho DF, Sant SV, Shin H, Oliveira JT, Gomes ME, Neves NM, Khademhosseini A, Reis RL (2010) Modified Gellan Gum hydrogels with tunable physical and mechanical properties. Biomaterials 31(29):7494–7502

27. Al-Ahmady Z, Kostarelos K (2016) Chemical components for the design of temperature-responsive vesicles as cancer therapeutics. Chem Rev 116(6):3883–3918

28. Roy D, Brooks WL, Sumerlin BS (2013) New directions in thermoresponsive polymers. Chem Soc Rev 42(17):7214–7243

29. Brun-Graeppi AKAS, Richard C, Bessodes M, Scherman D, Merten O-W (2010) Thermoresponsive surfaces for cell culture and enzyme-free cell detachment. Prog Polym Sci 35(11):1311–1324

30. Shimoboji T, Larenas E, Fowler T, Kulkarni S, Hoffman AS, Stayton PS (2002) Photoresponsive polymer-enzyme switches. Proc Natl Acad Sci USA 99(26):16592–16596

31. Jochum FD, Theato P (2013) Temperature- and light-responsive smart polymer materials. Chem Soc Rev 42(17):7468–7483

32. Zhuang J, Gordon MR, Ventura J, Li L, Thayumanavan S (2013) Multi-stimuli responsive macromolecules and their assemblies. Chem Soc Rev 42(17):7421–7435

33. Katayama Y, Sonoda T, Maeda M (2001) A polymer micelle responding to the protein kinase a signal. Macromolecules 34(24):8569–8573

34. Sonoda T, Nogami T, Oishi J, Murata M, Niidome T, Katayama Y (2005) A peptide sequence controls the physical properties of nanoparticles formed by peptide–polymer conjugates that respond to a protein kinase A signal. Bioconjug Chem 16(6):1542–1546

35. Beltran S, Baker JP, Hooper HH, Blanch HW, Prausnitz JM (1991) Swelling equilibria for weakly ionizable, temperature-sensitive hydrogels. Macromolecules 24(2):549–551

36. Aoyagi T, Ebara M, Sakai K, Sakurai Y, Okano T (2000) Novel bifunctional polymer with reactivity and temperature sensitivity. J Biomat Sci-Polym E 11(1):101–110

37. Qiao SL, Wang Y, Lin YX, An HW, Ma Y, Li LL, Wang L, Wang H (2016) Thermo-controlled in situ phase transition of polymer-peptides on cell surfaces for high-performance proliferative inhibition. ACS Appl Mat Inter 8(27):17016–17022

38. Blanazs A, Armes SP, Ryan AJ (2009) Self-assembled block copolymer aggregates: from micelles to vesicles and their biological applications. Macromol Rapid Commun 30(4–5):267–277

39. Qin S, Geng Y, Discher DE, Yang S (2006) Temperature-controlled assembly and release from polymer vesicles of Poly(ethylene oxide)-block- poly(N-isopropylacrylamide). Adv Mater 18(21):2905–2909

40. de Graaf AJ, Boere KW, Kemmink J, Fokkink RG, van Nostrum CF, Rijkers DT, van der Gucht J, Wienk H, Baldus M, Mastrobattista E, Vermonden T, Hennink WE (2011) Looped structure of flowerlike micelles revealed by 1H NMR relaxometry and light scattering. Langmuir 27(16):9843–9848

41. Li Y, Lokitz BS, McCormick CL (2006) Thermally responsive vesicles and their structural "locking" through polyelectrolyte complex formation. Angew Chem Int Ed 45(35):5792–5795

42. Grubbs RB, Sun Z (2013) Shape-changing polymer assemblies. Chem Soc Rev 42(17):7436–7445

43. Fenske T, Korth HG, Mohr A, Schmuck C (2012) Advances in switchable supramolecular nanoassemblies. Chemistry 18(3):738–755

44. Zhao Y, Sakai F, Su L, Liu Y, Wei K, Chen G, Jiang M (2013) Progressive macromolecular self-assembly: from biomimetic chemistry to bio-inspired materials. Adv Mater 25(37):5215–5256

45. Shimizu T, Masuda M, Minamikawa H (2005) Supramolecular nanotube architectures based on amphiphilic molecules. Chem Rev 105(4):1401–1443

46. Sundararaman A, Stephan T, Grubbs RB (2008) Reversible restructuring of aqueous block copolymer assemblies through stimulus-induced changes in amphiphilicity. J Am Chem Soc 130(37):12264

47. Moughton AO, O'Reilly RK (2010) Thermally induced micelle to vesicle morphology transition for a charged chain end diblock copolymer. Chem Commun 46(7):1091–1093

48. Moughton AO, Patterson JP, O'Reilly RK (2011) Reversible morphological switching of nanostructures in solution. Chem Commun 47(1):355–357

49. Cai Y, Aubrecht KB, Grubbs RB (2011) Thermally induced changes in amphiphilicity drive reversible restructuring of assemblies of ABC triblock copolymers with statistical polyether blocks. J Am Chem Soc 133(4):1058–1065

50. Ding T, Valev VK, Salmon AR, Forman CJ, Smoukov SK, Scherman OA, Frenkel D, Baumberg JJ (2016) Light-induced actuating nanotransducers. Proc Natl Acad Sci 113 (20):5503–5507

51. Wei K, Su L, Chen G, Jiang M (2011) Does PNIPAM block really retard the micelle-to-vesicle transition of its copolymer? Polymer 52(16):3647–3654

52. Liu X, Liu Y, Zhang Z, Huang F, Tao Q, Ma R, An Y, Shi L (2013) Temperature-responsive mixed-shell polymeric micelles for the refolding of thermally denatured proteins. Chemistry 19(23):7437–7442

53. Huang F, Wang J, Qu A, Shen L, Liu J, Liu J, Zhang Z, An Y, Shi L (2014) Maintenance of amyloid beta peptide homeostasis by artificial chaperones based on mixed-shell polymeric micelles. Angew Chem Int Ed 53(34):8985–8990

54. Gao H, Xiong J, Cheng T, Liu J, Chu L, Liu J, Ma R, Shi L (2013) In vivo biodistribution of mixed shell micelles with tunable hydrophilic/hydrophobic surface. Biomacromol 14(2):460–467

55. Uchiyama S, Matsumura Y, de Silva AP, Iwai K (2003) Fluorescent molecular thermometers based on polymers showing temperature-induced phase transitions and labeled with polarity-responsive Benzofurazans. Anal Chem 75(21):5926–5935

56. Gota C, Okabe K, Funatsu T, Harada Y, Uchiyama S (2009) Hydrophilic fluorescent nanogel thermometer for intracellular thermometry. J Am Chem Soc 131(8):2766–2767

57. Okabe K, Inada N, Gota C, Harada Y, Funatsu T, Uchiyama S (2012) Intracellular temperature mapping with a fluorescent polymeric thermometer and fluorescence lifetime imaging microscopy. Nat commun 28(3):705

58. Malmsten M, Lindman B (1992) Self-assembly in aqueous block copolymer solutions. Macromolecules 25(20):5440–5445

59. Jeong B, Gutowska A (2002) Lessons from nature: stimuli-responsive polymers and their biomedical applications. Trends Biotechnol 20(7):305–311
60. Mortensen K, Pedersen JS (1993) Structural study on the micelle formation of poly(ethylene oxide)-poly(propylene oxide)-poly(ethylene oxide) triblock copolymer in aqueous solution. Macromolecules 26(4):805–812
61. Lutz J-F (2011) Thermo-switchable materials prepared using the OEGMA-platform. Adv Mater 23(19):2237–2243
62. Lim YH, Tiemann KM, Heo GS, Wagers PO, Rezenom YH, Zhang S, Zhang F, Youngs WJ, Hunstad DA, Wooley KL (2015) Preparation and in vitro antimicrobial activity of silver-bearing degradable polymeric nanoparticles of polyphosphoester-block-Poly (l-lactide). ACS Nano 9(2):1995–2008
63. Yilmaz ZE, Jerome C (2016) Polyphosphoesters: new trends in synthesis and drug delivery applications. Macromol Biosci 16(12):1745–1761
64. Matsumoto NM, Prabhakaran P, Rome LH, Maynard HD (2013) Smart vaults: thermally-responsive protein nanocapsules. ACS Nano 7(1):867–874
65. Lin EW, Boehnke N, Maynard HD (2014) Protein-polymer conjugation via ligand affinity and photoactivation of glutathione S-transferase. Bioconjug Chem 25(10):1902–1909
66. Lin E-W, Maynard HD (2015) Grafting from small interfering ribonucleic acid (siRNA) as an alternative synthesis route to siRNA–polymer conjugates. Macromolecules 48(16):5640–5647
67. Vazquez-Dorbatt V, Lee J, Lin EW, Maynard HD (2012) Synthesis of glycopolymers by controlled radical polymerization techniques and their applications. Chembiochem Eur J Chem Bio 13(17):2478–2487
68. Wong CK, Laos AJ, Soeriyadi AH, Wiedenmann J, Curmi PM, Gooding JJ, Marquis CP, Stenzel MH, Thordarson P (2015) Polymersomes prepared from thermoresponsive fluorescent protein-polymer bioconjugates: capture of and report on drug and protein payloads. Angew Chem Int Ed 54(18):5317–5322
69. Qiao S-L, Ma Y, Wang Y, Lin Y-X, An H-W, Li L-L, Wang H (2017) General approach of stimuli-induced aggregation for monitoring tumor therapy. ACS Nano 11(7):7301–7311
70. Goodall S, Howard CB, Jones ML, Munro T, Jia Z, Monteiro MJ, Mahler S (2015) An EGFR targeting nanoparticle self assembled from a thermoresponsive polymer. J Chem Technol Biot 90(7):1222–1229
71. Lynn GM, Laga R, Darrah PA, Ishizuka AS, Balaci AJ, Dulcey AE, Pechar M, Pola R, Gerner MY, Yamamoto A, Buechler CR, Quinn KM, Smelkinson MG, Vanek O, Cawood R, Hills T, Vasalatiy O, Kastenmuller K, Francica JR, Stutts L, Tom JK, Ryu KA, Esser-Kahn AP, Etrych T, Fisher KD, Seymour LW, Seder RA (2015) In vivo characterization of the physicochemical properties of polymer-linked TLR agonists that enhance vaccine immunogenicity. Nat Biotechnol 33(11):1201–1210
72. Liu Y, Han X, He L, Yin Y (2012) Thermoresponsive assembly of charged gold nanoparticles and their reversible tuning of plasmon coupling. Angew Chem Int Ed 51(26):6373–6377
73. Meyerbroker N, Kriesche T, Zharnikov M (2013) Novel ultrathin poly(ethylene glycol) films as flexible platform for biological applications and plasmonics. ACS Appl Mater Interfaces 5 (7):2641–2649
74. Lim S, Song JE, La JA, Cho EC (2014) Gold nanospheres assembled on hydrogel colloids display a wide range of thermoreversible changes in optical bandwidth for various plasmonic-based color switches. Chem Mater 26(10):3272–3279
75. Ieong NS, Brebis K, Daniel LE, O'Reilly RK, Gibson MI (2011) The critical importance of size on thermoresponsive nanoparticle transition temperatures: gold and micelle-based polymer nanoparticles. Chem Commun 47(42):11627–11629

76. Gibson MI, Paripovic D, Klok HA (2010) Size-dependent LCST transitions of polymer-coated gold nanoparticles: cooperative aggregation and surface assembly. Adv Mater 22(42):4721–4725
77. Won S, Phillips DJ, Walker M, Gibson MI (2016) Co-operative transitions of responsive-polymer coated gold nanoparticles; precision tuning and direct evidence for co-operative aggregation. J Mater Chem B 4(34):5673–5682

# Chapter 7
# Self-assembled Nanomaterials for Autophagy Detection and Enhanced Cancer Therapy Through Modulating Autophagy

Yao-Xin Lin, Yi Wang and Hao Wang

**Abstract** Autophagy is a highly conserved cellular metabolism process and has become a diagnostic and therapeutic target for some diseases. In recent years, the applications of excellent self-assembled nanosystems in autophagy detection and modulation increase rapidly and provide promising approaches for disease diagnostics and therapeutics. Here, we review the recent advances in self-assembled nanomaterials for autophagy detection and autophagy-mediated cancer therapy. Meanwhile, the nanoprobe based on "in vivo self-assembly" strategy is highlighted. The "in vivo self-assembly" endows nanoprobes with higher accumulation, longer and better signal stability for in vivo bioimaging. Next, we will introduce the progress of self-assembled nanomaterials as autophagy modulators or drug carriers for enhanced cancer therapy through autophagy modulation.

**Keywords** Self-assembly · Nanomaterials · Autophagy · Imaging
Cancer therapy

Y.-X. Lin · Y. Wang · H. Wang (✉)
CAS Center for Excellence in Nanoscience, CAS Key Laboratory for Biomedical Effects of Nanomaterials and Nanosafety, National Center for Nanoscience and Technology (NCNST), Beijing 100190, People's Republic of China
e-mail: wanghao@nanoctr.cn

Y.-X. Lin
e-mail: linyx@nanoctr.cn

Y. Wang
e-mail: wangyi@nanoctr.cn

Y.-X. Lin · Y. Wang · H. Wang
University of Chinese Academy of Sciences, Beijing 100049, People's Republic of China

© Springer Nature Singapore Pte Ltd. 2018
H. Wang and L.-L. Li (eds.), *In Vivo Self-Assembly Nanotechnology for Biomedical Applications*, Nanomedicine and Nanotoxicology,
https://doi.org/10.1007/978-981-10-6913-0_7

## 7.1 Introduction

Autophagy is a highly conserved cellular metabolism process whereby damaged organelles and misfolded proteins are degraded using double-membraned autophagosomes [1]. Three main types of autophagy are described, i.e., chaperone-mediated autophagy, micro-autophagy, and macro-autophagy. In this chapter, the term "autophagy" refers to macro-autophagy, which is the best understood of these. During the autophagy process, long-lived proteins and damaged organelles are first captured by double-bilayer membranes to form the autophagosome and then are degraded upon fusion with lysosome [1–3]. So, autophagy is a dynamic process, which includes the initial formation, elongation, and maturation of autophagosome, and finally fusion with lysosome to degrade by lysosomal enzymes (Fig. 7.1). Autophagic degradation process maintains cellular homoeostasis and promotes survival in almost cell because intracellular components are recycled and serve as an alternative source of energy [4]. Moreover, this fundamental degradation mechanism is more important to the cells that are under pathophysiological conditions, including starvation (e.g., growth factor/nutrient deprivation), oxidative stress, aging, infection [5, 6]. Additionally, the dysfunction of autophagy has close correlations with various human diseases [7, 8] such as neurodegenerative, autoimmune, infectious diseases [9], cancer [10–12], aging [13]. Therefore, autophagy has been an important diagnostic and therapeutic target for various diseases [14].

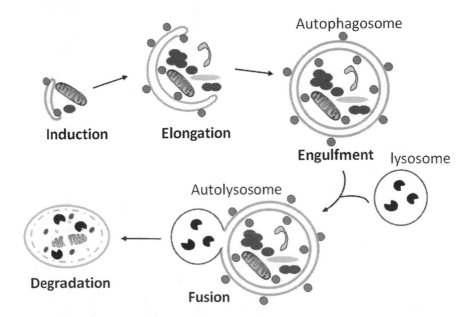

**Fig. 7.1** Schematic illustration of autophagic process

Nanomaterials that have diameter between 10 and 200 nm of at least one dimension are widely used in biological imaging and drug delivery field [15–17] due to their excellent physical and chemical characteristics, such as controlled size, large surface area-to-volume ratio, facile surface modification, and good biocompatibility. Compared to simple molecular probes, the nanomaterials-based probes have unique optical, electronic, acoustic, magnetic, and chemical properties [16]. Additionally, nanostructure-based drug system incorporates specific features, such as long circulation and reduced toxicity, and could selectively permeate and accumulate in solid tumors via the well-known enhanced permeability and retention (EPR) effect due to its nanoscale size. Meanwhile, the nanomaterials can be functionalized with multiple groups, such as specific cell targets for active targeting, biocompatible surface coatings for modulating pharmacokinetics and improving bio-distribution, and co-delivering imaging and therapeutic agents for disease theranostics. Besides, some nanomaterials have closely correctively with the autophagy, and meanwhile, autophagy shows strongly positive outcome for cancer therapies. Therefore, studying engineered nanomaterials for autophagy-mediated cancer theranostics is an emerging field of tumor research.

In this review, we aim to present a discussion on the self-assembled nanoma terials for autophagy detection and enhanced cancer therapy through modulating autophagy. Firstly, we introduce recent advancements in the development of nanostructure-based probes for autophagy detection. We here summarize the structures and imaging mechanisms of these nanoprobes and briefly discuss their potential applications in autophagy-mediated diseases. In addition, we will highlight the use of the "in vivo self-assembly" strategy for real-time and quantitative monitoring autophagy in vivo. Secondly, we focus on recent examples of self-assembled nanomaterials for autophagy-mediated cancer therapy.

## 7.2 Self-assembled Nanostructure-Based Probes for Autophagy Detection

Nowadays, several approaches are utilized to evaluate autophagy, e.g., transmission electron microscopy (TEM), Western blot, immunofluorescence of LC3 [18]. TEM could be used for detection of autophagy by morphological analysis of autophagic structures and has been widely recognized as a gold standard for autophagy detection, but it needs several days for sampling artifact and it cannot be applied to living specimen for the real-time monitoring of studies. GFP-LC3/Atg8, a fusion of the autophagosome marker LC3 (microtubule-associated protein 1 light chain 3) and GFP (green fluorescent protein), is widely used as a reliable probe for detection of autophagy, but with limitations such as time-consuming on probe construction, large transfected protein may have negative influence on the physiological status of the cells and unable to long-time monitor because the GFP fluorescence in autolysosome is attenuated due to the degradation and quenching in the acidic environment. Besides, it does not distinguish between "autophagy active" and "autophagy inactive" states. Western blot on LC3 protein is a semiquantitative method to evaluate autophagy, but there is an issue with the limited dynamic range of LC3 blots and it cannot be applied to living cells, either.

### 7.2.1 Ex Situ Self-assembled Nanoprobes

Environment-triggered nanomaterials that are highly responsive to the polarity [19], temperature [20], pH [21, 22], and enzyme of the microenvironment [23–26] have been widely used in biological imaging [27, 28] and drug delivery fields [29, 30]. Among them, the self-assembled nanoprobes with "turn-on" signals for tumor diagnostics under specifically physiological or pathological environments are booming and grabbing a great deal of attention. ATG4B is one of the most important and indispensable autophagy-related cysteine proteases [31, 32] and thereby has been utilized to act as a potential autophagic biomarker [33–36]. Based on these facts, some molecular and nanoprobes for ATG4B detection were developed. For example, Choi et al. described a peptide-conjugated polymeric nanoparticle [33] which was used for real-time visualization ATG4 activity in living cells. In this study, the peptide (STFGFSGKRRRRRRRRR) that was highly sensitive to the ATG4B was used as a switch of the nanoprobes, and the fluorescent dye FITC and the quencher BHQ1 were acted as signal units modified on the peptide to form FITC-peptide-conjugated complex, which ultimately self-assembled into stable nanoparticles (TFG-HGC) with "turn-off" signals due to fluorescence resonance energy transfer (FRET) effect. Once upon incubation with ATG4B, it would switch on the enhanced fluorescence. Meanwhile, a lysosome-staining dye (Lycolite Red) was captured into the hydrophobic core of the nanoparticles, which acted another signal motif to distinguish the ATG4B enzymatic reaction from nonspecific lysosomal degradation (Fig. 7.2). The cellular experimental results demonstrated that there were "turn-on" signals in autophagy-induced cells ("autophagy active"), not autophagy-inhibited cells ("autophagy inactive"). Moreover, this nanoprobe could be used for real-time monitoring ATG4B activity in living cells and offered quantitative results.

### 7.2.2 In Situ Constructed Nanoprobe Based on "In Vivo Self-assembly" Strategy

Many concerns about the signal stability, long-time imaging, and safety of nanoprobes should be assessed [16, 37] before the clinical applications. Therefore, an "in vivo self-assembly" strategy was described by Wang group [38], i.e., in situ controllable construction of functional nanostructures from molecules or

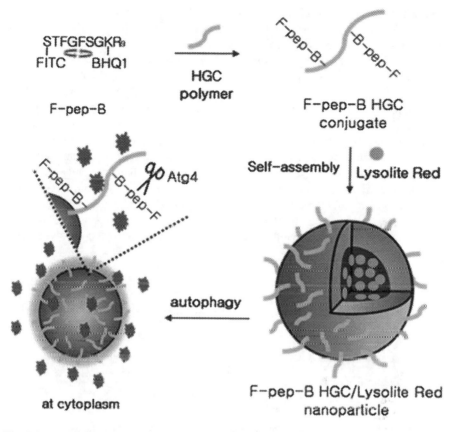

**Fig. 7.2** Schematic representation of preparation and autophagy-imaging mechanism of the peptide-conjugated polymeric nanoparticle

nanoparticles in biological/pathological environment for desired targeting, accumulation, retention. Based on this concept, they prepared a series of responsive molecules/nanoparticles that could simultaneously self-assemble into nanostructures in targeting site for in vivo biomedical imaging and tumor treatment [38–43]. Among them, a "molecule-like" responsive nanoprobe that could in situ self-assemble into new nanostructures in living cells with "turn-on" signal was prepared for autophagy detection [43]. The nanoprobe consisted of three modular functional motifs, i.e., a peptide (GTFGFSGKG) that was responsive to Atg4B as a responsive unit, a bis(pyrene) derivative (BP) with aggregation-induced emission (AIE) characteristic as signal molecule, and a bulky hydrophilic dendrimer (PAMAM) as a carrier which increased the hydrophilic of probes and ensured monomeric signal molecule in saline/normal physiological environment. Therefore, the probe maintained the quenched fluorescence in water solution, but it would emit enhanced fluorescence upon Atg4B incubation, and its enhanced fluorescence intensity (525 nm) was correlated with Atg4B concentration in a linear manner.

Meanwhile, transmission electron microscopy (TEM) images showed the nanoprobes finally self-assembled into fiber-like nanostructures after Atg4B incubation. The results of living cells images demonstrated that this nanoprobe was successfully utilized for real-time monitoring autophagy. However, the optical properties of the probes in the visible range of the spectrum limited its application in highly living organisms. To solve this problem, a photoacoustic (PA) nanoprobe based on "in vivo self-assembly" strategy for real-time and quantitative detection of autophagy in mice was developed [44]. Photoacoustic (PA) tomography was a noninvasive bioimaging technique and be widely used for in vivo disease diagnosis [45–47]. Using the photoacoustic nanoprobe, the real-time imaging data of autophagy of tumors was obtained; therein, it was employed for enhancing the efficacy of autophagy-mediated chemotherapy and minimized the toxicity through regulation of autophagy during chemotherapy (Fig. 7.3). Above all, "in vivo self-assembly" strategy showed rapid, real-time properties, without tedious sample preparation, and was thus believed to be an efficient way to determine autophagy in vitro and in vivo.

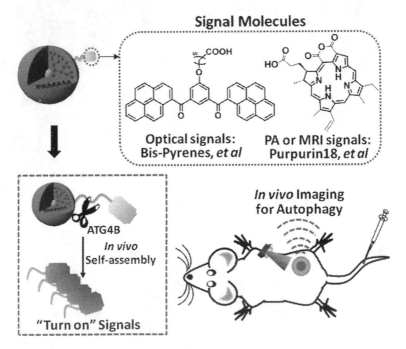

**Fig. 7.3** Schematic illustration of nanoprobe based on "in vivo self-assembly" strategy for real-time and quantitative detection of autophagy

## 7.3  Self-assembled Nanomaterials Modulate Autophagy in Cancer Therapy

### 7.3.1  Role of Autophagy in Cancer and Cancer Therapy

In tumor cells, autophagy is complex and plays a dual role in different tumor cells and different stages of tumorigenesis [11, 12, 48]. On the one hand, autophagy is a tumor-suppression mechanism for inhibiting tumor cell growth. For example, the autophagy gene Beclin-1 (*BECN1*, ATG 6 gene in yeast) [49, 50] is an important and essential part of type III phosphatidylinositol 3-kinase (PI3K) complex, which regulates autophagosome formation, but it is monoallelically lost in human breast, ovarian cancer, etc. Increasing evidence has indicated that *BECN1* can be as a tumor suppressor in the early stages of tumorigenesis. In addition, excessive and prolonged autophagy will promote apoptosis and lead to cell death, which is known as autophagic cell death or programmed cell death type II [51, 52]. On the other hand, autophagy serves as a pro-survival process for clearing proteins and organelles. In comparison with normal cells, the cancer cells also rely on autophagic degradation process. In some stress stimuli, such as nutrient deprivation hypoxia and ROS generation, cancer cells would upregulate autophagy in order for them to resist these stresses [4]. During chemo-, radio-photodynamic-therapeutic treatments, autophagy is induced and acts as a protective mechanism of cancer cells for them to survive [53].

Taken together, autophagy functions both as a pro-survival mechanism that helps tumor cells resist death and metabolic stresses triggered by therapeutic agents and as a tumor suppressor that promotes tumor cells death (Fig. 7.4). Therefore, we need to confirm the autophagic affects in tumor development, progression, and treatment, as to use this information to correctly modulate autophagy and enhance cancer therapy.

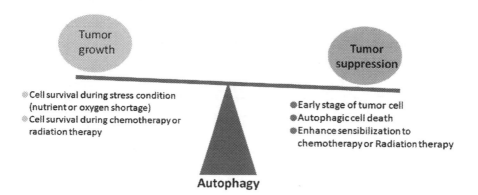

**Fig. 7.4**  Dual roles of autophagy in cancer cells

### 7.3.2 The Chemical Agents for Autophagy Induction or Inhibition

Nowadays, there are two types of agents for autophagy modulation [54], i.e., autophagy inducers and autophagy inhibitors (Fig. 7.5). In mammalian cells, the mTOR signaling pathway [55] regulates cell growth and proliferation, protein synthesis and degradation, etc. Some upstream signals [56], such as growth factors, amino acids, and glucose, are integrated by mTOR. Thus, the inhibition of mTOR complex1 (mTORC1) that results from mTOR inhibitors induces autophagy through acting on ULK1 (*ATG1*, autophagic gene) complex. For example, rapamycin and its analogs (i.e., everolimus (RAD-001)) are mTOR inhibitors and could stimulate autophagy activation. Additional, some anticancer agents could also induce autophagy. For example, tamoxifen, a nonsteroidal estrogen receptor (ER) antagonist, has been widely used as a chemotherapeutic agent against breast cancer, but could induce cell death by an autophagic mechanism.

Autophagy inhibitors can be broadly classified as early-stage versus late-stage inhibitors of the pathway. Early-stage inhibitors consist of 3-methyladenine (3-MA), wortmannin, and so on, while late-stage inhibitors include chloroquine (CQ) or hydroxychloroquine (HCQ), and bafilomycin A1. CQ or HCQ is lysosome drugs that upregulate the acid environment of lysosome resulting in damage of

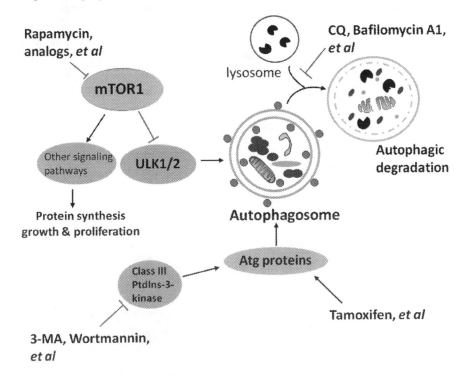

**Fig. 7.5** Chemical agents for modulation of autophagy

autophagic degradation. Bafilomycin A1 is an ATPase inhibitor that prevents fusion of autophagosomes with the lysosomes. In addition, many other specific inhibitors of the autophagic machinery (e.g., siRNA of *Atg* gene, PIK3 inhibitors, etc.) show potential use in autophagy regulation.

## 7.3.3  Self-assembled Nanomaterials Modulate Autophagy in Cancer Therapy

Given the paradoxical role of autophagy in cancer and cancer therapy, it is not surprising that modulation of autophagy to promote tumor cell death would be beneficial method for improving therapy. However, the most of autophagy-modulating drugs, such as rapamycin and bafilomycin A1, are hydrophobic so that hardly to be directly used in clinical application. Recent progress in self-assembled nanomaterials has deeply impacted the field of drug delivery. On the one hand, physicochemical characters of nanostructures, including small size, high loading capacity, and reversible character, are major assets to the delivery of chemotherapeutics and/or biologics to solid tumors. On the other hand, nanomaterials are not only capable of delivering multiple and highly concentrated therapeutic payloads, but also simultaneously release them in the micro-environment of tumor. In addition, some biomolecules are used to modify the surface of the self-assembled nanostructures that are of high interest to confer them on good biocompatibility and long circulation. Currently, the most widely used nanocarrier systems are liposome, micelle, polymeric nanoparticles (Fig. 7.6). All these nanocarriers composed of biomolecules, which could enhance the drugs water solution, reduce systemic side effects, and increase therapeutic efficiency.

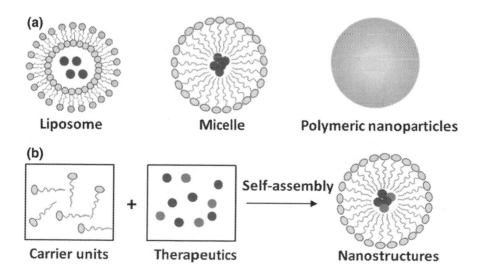

**Fig. 7.6** Examples of nanocarrier for drug delivery

### 7.3.3.1 Self-assembled Nanomaterials for Therapeutics/ Autophagy-Modulating Drugs Delivery

A single payload of drug delivery: Rapamycin is an autophagy inducer and acts as potential anticancer drug for the treatment of rapamycin-sensitive cancers, but its poor water solubility hampers the application to in vivo. Therefore, Chen et al. [57] designed a self-assembled micelles that consisted of mPEG-b-P (HPMA-Lac-co-His) (poly(ethylene lycol)-b-poly(N-(2-hydroxypropyl) methacrylamide dilactate)-co-(N-(2-hydroxypropyl) methacrylamide-co-histidine)) and mPEG-b-PLA(poly(ethylene glycol)-b-poly(D,L-lactide)) diblock copolymers. The micelles not only improved the solubility of rapamycin but also enhanced intracellular delivery for increasing antitumor efficacy. Meanwhile, the micelles contained a pH-sensitive structural feature, which maintained micelles nanostructures under the neutral environment, but was disruptive in intracellular acid organelles, resulting in the rapid release of the drugs. Confocal imaging results clearly revealed that encapsulated rapamycin was released at the lysosome and induced autophagolysosome formation. In vivo tumor growth inhibition results showed that the rapamycin-loaded micelles exhibited excellent antitumor activity.

Besides, there were some nanomaterials that only delivery therapeutics and the autophagy-modulating agents needed to be additional used. For example, Hou et al. developed a novel self-assembled nanoparticle (Fig. 7.7) for the delivery of mRIP3 pDNA (receptor-interacting protein kinase 3) [58]. RIP3 is a member of the RIP kinase family, which plays a key role in necroptosis. CQ could significantly upregulate RIP3 expression, while RIP3 is also involved in CQ-related autophagy. In this study, the authors found that using mRIP3-loaded nanoparticles in combination with CQ induced significant tumor death, indicating a synergistic antitumor effect through modulating autophagy.

Multiple agents' delivery: Increasing evidence has demonstrated that a single therapeutic treatment (e.g., chemotherapy) for cancer therapy is unsatisfactory. The reason may be that the chemotherapeutics could induce autophagy of tumor cells, which serves as pro-survival mechanism to enhance the resistance of tumor cells to chemotherapy. Thus, co-delivery of chemotherapeutics and autophagy inhibitors is a promising approach to increase the efficiency of cancer therapy. Given this, numerous efforts have been made in recent years to develop combined treatments based on nanotechnology [59–66]. Among them, the self-assembled nanomaterials that have the capacity to co-load chemotherapeutics (e.g., cisplatin, doxorubicin) and autophagy inhibitors (e.g., CQ/HCQ, siRNA of *Atg* gene, etc.) for synergetic treatment draw attracting attention in the field of cancer therapy. For example, Jia and his colleagues reported a self-assembled polymeric nanoparticle with pH sensitivity to co-deliver *Beclin1* siRNA and DOX for combinational tumor treatments [64]. In this study (Fig. 7.8), phenylboronic acid-tethered hyperbranched oligoethylenimine (OEI600-PBA) and 1,3-diol-rich hyperbranched polyglycerol (HBPO) were prepared and self-assembled into a core-corona nanostructure. Therein, the inner core of nanoparticles captured anticancer drugs through hydrophobic interaction and the surface cationic OEI600-PBA units absorbed

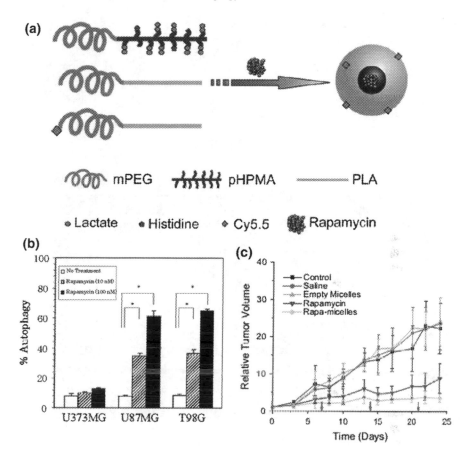

**Fig. 7.7** Self-assembled micelles for a single payload of drug delivery

siRNA. It was demonstrated that the nanoparticles could silence the Becline1 expression, thereby suppress the DOX-induced autophagy, and consequently provide strong synergism with a significant enhancement of cell-killing effects on tumor cells.

Photothermal therapy (PTT) was a promising strategy for cancer therapy, but it also induced autophagy of cancer cells. Inhibition of autophagy of tumor cells would significantly enhance the efficacy of photothermal killing of cancer cells. Thus, Zhou and his co-workers described a chloroquine-loaded polydopamine nanoparticle for sensitized photothermal cancer therapy [67]. In vitro and in vivo results demonstrated that inhibition of autophagy remarkably increased the efficacy of photothermal therapy, leading to efficient tumor suppression at a mild temperature.

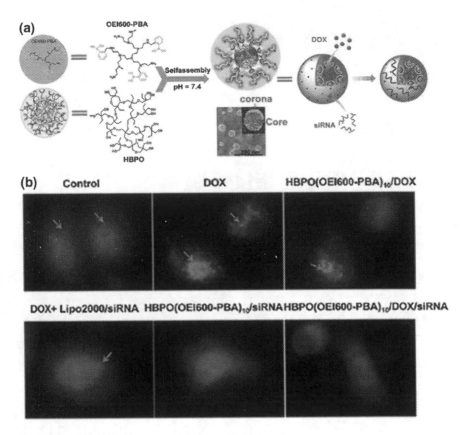

**Fig. 7.8** Self-assembled nanomaterials for multiple drug delivery

### 7.3.3.2 Self-assembled Nanomaterials for Synergistic Autophagy Induction

The above nanomaterials showed potential application in modulating autophagy by loading autophagy inducers or inhibitors, but the nanomaterials served as carrier had no effect on autophagy. Increasing evidence demonstrated that nanoparticles functioned as positive or negative agents in the process of autophagy. Therefore, using nanomaterials served as modulators of autophagy in combination with other molecules, such as chemotherapeutics and autophagy-modulating drugs, was of growing interest in autophagy-mediated cancer therapy. Given this, Wang and her co-workers described a self-assembled pH-sensitive polymeric nanoparticle (Fig. 7.9) with autophagy-inducing peptide for highly efficient induction of autophagy and enhanced treatment of breast cancer [68]. In this study, Beclin1, an autophagy-inducing peptide, was covalently grafted onto pH-sensitive polymers. The pH-sensitive polymers could induce autophagy through impacting lysosomes, and highly concentrated nanocarrier polymers led to excessive impairment of lysosomes, which blocked autophagic flux and ultimately resulted in autophagic cell death [69]. The synergistic autophagy-inducing effects of polymers and Beclin1 peptides effectively induced autophagy and effectively inhibited the growth of breast tumors in vivo. In addition, based on similar strategy, the authors designed pH-sensitive polymeric nanoparticles with gold (I) compound payloads for synergistically induced cancer cell death through regulation of autophagy [70]. All these results demonstrated that identification of the nanomaterials combined with other agents for synergistically inducing tumor cell death through regulating autophagy may open a new avenue for cancer treatment.

## 7.4 Conclusion and Outlook

In conclusion, autophagy exhibits a tight relationship with various human diseases and the exploitation of the autophagy detection and modulation by nanomaterials as a new theranostic strategy is under investigation. We find researchers have developed various responsive nanoprobes for in vivo autophagy-imaging and designed numerous smart nanoassemblies for cancer therapy through modulating autophagy.

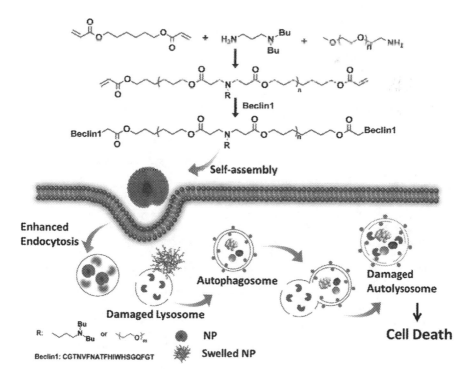

**Fig. 7.9** Self-assembled nanomaterials for autophagy-mediated therapy through therapeutics delivery and autophagy induction

However, there are some more challenges to be faced for their clinical applications. For example, the molecular mechanism of how autophagy inhibits tumor development and promotes cell death remains unclear. Autophagy plays a complex role in cancer cells and cancer therapy. Both of autophagy induction and inhibition may provide beneficial effects for cancer therapy. Hence, it would be of great importance to obtain autophagic information during treatment. We here highlight several nanostructures-based probes for autophagy detection. These innovative nanoprobes could overcome limitations of conventional detecting approaches and are capable of in vivo real-time and quantitative detection of autophagy. Moreover, according to the results of imaging, we could design and optimize therapeutic protocol through precisely modulating autophagy. In addition, increasing evidence indicate that self-assembled nanomaterials have autophagy-modulating effects on tumor cells, i.e., autophagy induction or inhibition, which result in different biological consequence. Thus, understanding the synergy between autophagy, cancer therapy, and nanomaterials could be pivotal to design new therapeutic cancer strategies.

**Acknowledgements** This work was supported by the National Basic Research Program of China (973 Program, 2013CB932701), National Natural Science Foundation of China (21374026, 51573031 and 51573032), Science Fund for Creative Research Groups of the National Natural Science Foundation of China (11621505), Key Project of Chinese Academy of Sciences in Cooperation with Foreign Enterprises (GJHZ1541).

# References

1. Mizushima N (2007) Autophagy: process and function. Genes Dev 21(22):2861–2873
2. Xie Z, Klionsky DJ (2007) Autophagosome formation: core machinery and adaptations. Nat Cell Biol 9(10):1102–1109
3. Yang ZF, Klionsky DJ (2010) Eaten alive: a history of macroautophagy. Nat Cell Biol 12 (9):814–822
4. Green DR, Levine B (2014) To be or not to be? How selective autophagy and cell death govern cell fate. Cell 157(1):65–75
5. Klionsky DJ, Emr SD (2000) Cell biology—autophagy as a regulated pathway of cellular degradation. Science 290(5497):1717–1721
6. Rabinowitz JD, White E (2010) Autophagy and metabolism. Science 330(6009):1344–1348
7. Levine B, Kroemer G (2008) Autophagy in the pathogenesis of disease. Cell 132(1):27–42
8. Rubinsztein DC, Codogno P, Levine B (2012) Autophagy modulation as a potential therapeutic target for diverse diseases. Nat Rev Drug Discovery 11(9):709–784
9. Kizilarslanoglu MC, Ulger Z (2015) Role of autophagy in the pathogenesis of Alzheimer disease. Turkish J Med Sci 45(5):998–1003
10. White E (2012) Deconvoluting the context-dependent role for autophagy in cancer. Nat Rev Cancer 12(6):401–410
11. Mathew R, Karantza-Wadsworth V, White E (2007) Role of autophagy in cancer. Nat Rev Cancer 7(12):961–967
12. Kondo Y, Kanzawa T, Sawaya R, Kondo S (2005) The role of autophagy in cancer development and response to therapy. Nat Rev Cancer 5(9):726–734
13. Rubinsztein DC, Marino G, Kroemer G (2011) Autophagy and aging. Cell 146(5):682–695

14. Rothe K, Lin H, Lin KBL, Leung A, Wang HM, Malekesmaeili M, Brinkman RR, Forrest DL, Gorski SM, Jiang X (2014) The core autophagy protein ATG4B is a potential biomarker and therapeutic target in CML stem/progenitor cells. Blood 123(23):3622–3634

15. Wang L, L-l Li, Y-s Fan, Wang H (2013) Host-guest supramolecular nanosystems for cancer diagnostics and therapeutics. Adv Mater 25(28):3888–3898

16. Smith BR, Gambhir SS (2017) Nanomaterials for in vivo imaging. Chem Rev 117(3):901–986

17. Hubbell JA, Chilkoti A (2012) Nanomaterials for drug delivery. Science 337(6092):303–305

18. Klionsky DJ, Abdalla FC, Abeliovich H, Abraham RT, Acevedo-Arozena A, Adeli K, Agholme L, Agnello M, Agostinis P et al (2012) Guidelines for the use and interpretation of assays for monitoring autophagy. Autophagy 8(4):445–544

19. Pandey S, Baker SN, Pandey S, Baker GA (2012) Fluorescent probe studies of polarity and solvation within room temperature ionic liquids. J Fluoresc 22(5):1313–1343

20. X-d Wang, Meier RJ, Schmittlein C, Schreml S, Schaeterling M, Wolfbeis OS (2015) A water-sprayable, thermogelating and biocompatible polymer host for use in fluorescent chemical sensing and imaging of oxygen, pH values and temperature. Sensor Actuat B Chem 221:37–44

21. Fan L, Gao S-Q, Li Z-B, Niu W-F, Zhang W-J, Shuang S-M, Dong C (2015) An indole-carbazole-based ratiometric emission pH fluorescent probe for imaging extreme acidity. Sensor Actuat B Chem 221:1069–1076

22. Wang L, Yang L, Cao D (2015) Probes based on diketopyrrolopyrrole and anthracenone conjugates with aggregation-induced emission characteristics for pH and BSA sensing. Sensor Actuat B Chem 221:155–166

23. Yuan Y, Sun HB, Ge SC, Wang MJ, Zhao HX, Wang L, An LN, Zhang J, Zhang HF, Hu B, Wang JF, Liang GL (2015) Controlled intracellular self-assembly and disassembly of F-19 nanoparticles for MR imaging of caspase 3/7 in zebrafish. ACS Nano 9(1):761–768

24. Hou Y, Zhou J, Gao ZY, Sun XY, Liu CY, Shangguan DH, Yang WS, Gao MY (2015) Protease-activated ratio metric fluorescent probe for pH mapping of malignant tumors. ACS Nano 9(3):3199–3205

25. Rivera-Gil P, De Koker S, De Geest BG, Parak WJ (2009) Intracellular processing of proteins mediated by biodegradable polyelectrolyte capsules. Nano Lett 9(12):4398–4402

26. Qiu LP, Zhang T, Jiang JH, Wu CC, Zhu GZ, You MX, Chen XG, Zhang LQ, Cui C, Yu RQ, Tan WH (2014) Cell membrane-anchored biosensors for real-time monitoring of the cellular microenvironment. J Am Chem Soc 136(38):13090–13093

27. Krueger AT, Kroll C, Sanchez E, Griffith LG, Imperiali B (2014) Tailoring chimeric ligands for studying and biasing erbb receptor family interactions. Angew Chem Int Ed 53(10):2662–2666

28. Yang Z, Cao J, He Y, Yang JH, Kim T, Peng X, Kim JS (2014) Macro/micro-environment-sensitive chemosensing and biological imaging. Chem Soc Rev 43(13):4563–4601

29. Mura S, Nicolas J, Couvreur P (2013) Stimuli-responsive nanocarriers for drug delivery. Nat Mater 12(11):991–1003

30. Stuart MAC, Huck WTS, Genzer J, Mueller M, Ober C, Stamm M, Sukhorukov GB, Szleifer I, Tsukruk VV, Urban M, Winnik F, Zauscher S, Luzinov I, Minko S (2010) Emerging applications of stimuli-responsive polymer materials. Nat Mater 9(2):101–113

31. Yu Z-Q, Ni T, Hong B, Wang H-Y, Jiang F-J, Zou S, Chen Y, Zheng X-L, Klionsky DJ, Liang Y, Xie Z (2012) Dual roles of Atg8-PE deconjugation by Atg4 in autophagy. Autophagy 8(6):883–892

32. Satoo K, Noda NN, Kumeta H, Fujioka Y, Mizushima N, Ohsumi Y, Inagaki F (2009) The structure of Atg4B-LC3 complex reveals the mechanism of LC3 processing and delipidation during autophagy. EMBO J 28(9):1341–1350

33. K-m Choi, Nam HY, Na JH, Kim SW, Kim SY, Kim K, Kwon IC, Ahn HJ (2011) A monitoring method for Atg4 activation in living cells using peptide-conjugated polymeric nanoparticles. Autophagy 7(9):1052–1062

34. Li M, Chen X, Ye Q-Z, Vogt A, Yin X-M (2012) A high-throughput FRET-based assay for determination of Atg4 activity. Autophagy 8(3):401–412
35. Ni Z, Gong Y, Dai X, Ding W, Wang B, Gong H, Qin L, Cheng P, Li S, Lian J, He F (2015) AU4S: A novel synthetic peptide to measure the activity of ATG4 in living cells. Autophagy 11(2):403–415
36. Vezenkov L, Honson NS, Kumar NS, Bosc D, Kovacic S, Nguyen TG, Pfeifer TA, Young RN (2015) Development of fluorescent peptide substrates and assays for the key autophagy-initiating cysteine protease enzyme, ATG4B. Bioorg Med Chem 23(13):3237–3247
37. Peynshaert K, Manshian BB, Joris F, Braeckmans K, De Smedt SC, Demeester J, Soenen SJ (2014) Exploiting intrinsic nanoparticle toxicity: the pros and cons of nanoparticle-induced autophagy in biomedical research. Chem Rev 114(15):7581–7609
38. Lin Y-X, Wang Y, Wang H (2017) Recent advances in nanotechnology for autophagy detection. Small 13:1700996
39. Zhang D, Qi G-B, Zhao Y-X, Qiao S-L, Yang C, Wang H (2015) In situ formation of nanofibers from purpurin18-peptide conjugates and the assembly induced retention effect in tumor sites. Adv Mater 27(40):6125–6130
40. Hu X-X, He P-P, Qi G-B, Gao Y-J, Lin Y-X, Yang C, Yang P-P, Hao H, Wang L, Wang H (2017) Transformable nanomaterials as an artificial extracellular matrix for inhibiting tumor invasion and metastasis. ACS Nano 11(4):4086–4096
41. Qiao S-L, Ma Y, Wang Y, Lin Y-X, An H-W, Li L-L, Wang H (2017) General approach of stimuli-induced aggregation for monitoring tumor therapy. ACS Nano 11(7):7301–7311
42. Qiao Z-Y, Lai W-J, Lin Y-X, Li D, Nan X-H, Wang Y, Wang H, Fang Q-J (2017) Polymer-KLAK peptide conjugates induce cancer cell death through synergistic effects of mitochondria damage and autophagy blockage. Bioconjug Chem 28(6):1709–1721
43. Lin Y-X, Qiao S-L, Wang Y, Zhang R-X, An H-W, Ma Y, Rajapaksha RPYJ, Qiao Z-Y, Wang L, Wang H (2017) An in situ intracellular self-assembly strategy for quantitatively and temporally monitoring autophagy. ACS Nano 11(2):1826–1839
44. Lin Y-X, Wang Y, Qiao S-L, An H-W, Wang J, Ma Y, Wang L, Wang H (2017) In vivo self-assembled nanoprobes for optimizing autophagy-mediated chemotherapy. Biomaterials 141:199–209
45. Zhang J, Qiao Z, Yang P, Pan J, Wang L, Wang H (2015) Recent advances in near-infrared absorption nanomaterials as photoacoustic contrast agents for biomedical imaging. Chin J Chem 33(1):35–52
46. Wang LV, Yao J (2016) A practical guide to photoacoustic tomography in the life sciences. Nat Meth 13(8):627–638
47. Taruttis A, Ntziachristos V (2015) Advances in real-time multispectral optoacoustic imaging and its applications. Nat Photon 9(4):219–227
48. Levine B (2007) Cell biology—autophagy and cancer. Nature 446(7137):745–747
49. Liang XH, Jackson S, Seaman M, Brown K, Kempkes B, Hibshoosh H, Levine B (1999) Induction of autophagy and inhibition of tumorigenesis by beclin 1. Nature 402(6762):672–676
50. Shoji-Kawata S, Sumpter R, Leveno M, Campbell GR, Zou ZJ, Kinch L, Wilkins AD, Sun QH, Pallauf K, MacDuff D, Huerta C, Virgin HW, Helms JB, Eerland R, Tooze SA, Xavier R, Lenschow DJ, Yamamoto A, King D, Lichtarge O, Grishin NV, Spector SA, Kaloyanova DV, Levine B (2013) Identification of a candidate therapeutic autophagy-inducing peptide. Nature 494(7436):201–206
51. Kroemer G, Levine B (2008) Autophagic cell death: the story of a misnomer. Nat Rev Mol Cell Biol 9(12):1004–1010
52. Denton D, Xu TQ, Kumar S (2015) Autophagy as a pro-death pathway. Immunol Cell Biol 93 (1):35–42
53. Gewirtz DA (2014) Autophagy and senescence in cancer therapy. J Cell Physiol 229(1):6–9
54. Fleming A, Noda T, Yoshimori T, Rubinsztein DC (2011) Chemical modulators of autophagy as biological probes and potential therapeutics. Nat Chem Biol 7(1):9–17

55. Laplante M, Sabatini DM (2012) mTOR signaling in growth control and disease. Cell 149 (2):274–293
56. Hay N, Sonenberg N (2004) Upstream and downstream of mTOR. Genes Dev 18(16):1926–1945
57. Chen Y-C, Lo C-L, Lin Y-F, Hsiue G-H (2013) Rapamycin encapsulated in dual-responsive micelles for cancer therapy. Biomaterials 34(4):1115–1127
58. Hou X, Yang C, Zhang L, Hu T, Sun D, Cao H, Yang F, Guo G, Gong C, Zhang X, Tong A, Li R, Zheng Y (2017) Killing colon cancer cells through PCD pathways by a novel hyaluronic acid-modified shell-core nanoparticle loaded with RIP3 in combination with chloroquine. Biomaterials 124:195–210
59. Sun R, Shen S, Zhang Y-J, Xu C-F, Cao Z-T, Wen L-P, Wang J (2016) Nanoparticle-facilitated autophagy inhibition promotes the efficacy of chemotherapeutics against breast cancer stem cells. Biomaterials 103:44–55
60. Zhang X, Dong Y, Zeng X, Liang X, Li X, Tao W, Chen H, Jiang Y, Mei L, Feng S-S (2014) The effect of autophagy inhibitors on drug delivery using biodegradable polymer nanoparticles in cancer treatment. Biomaterials 35(6):1932–1943
61. Zhang X, Zeng X, Liang X, Yang Y, Li X, Chen H, Huang L, Mei L, Feng S-S (2014) The chemotherapeutic potential of PEG-b-PLGA copolymer micelles that combine chloroquine as autophagy inhibitor and docetaxel as an anti-cancer drug. Biomaterials 35(33):9144–9154
62. Shi C, Zhang Z, Shi J, Wang F, Luan Y (2015) Co-delivery of docetaxel and chloroquine via PEO-PPO-PCL/TPGS micelles for overcoming multidrug resistance. Int J Pharm 495(2):932–939
63. Xu J, Zhu X, Qiu L (2016) Polyphosphazene vesicles for co-delivery of doxorubicin and chloroquine with enhanced anticancer efficacy by drug resistance reversal. Int J Pharm 498(1–2):70–81
64. Jia H-Z, Zhang W, Zhu J-Y, Yang B, Chen S, Chen G, Zhao Y-F, Feng J, Zhang X-Z (2015) Hyperbranched-hyperbranched polymeric nanoassembly to mediate controllable co-delivery of siRNA and drug for synergistic tumor therapy. J Control Release 216:9–17
65. Zhao L, Yang G, Shi Y, Su C, Chang J (2015) Co-delivery of Gefitinib and chloroquine by chitosan nanoparticles for overcoming the drug acquired resistance. J Nanobiotechnol 13
66. Babu A, Wang Q, Muralidharan R, Shanker M, Munshi A, Ramesh R (2014) Chitosan coated polylactic acid nanoparticle-mediated combinatorial delivery of cisplatin and siRNA/Plasmid DNA chemosensitizes cisplatin-resistant human ovarian cancer cells. Mol Pharm 11(8):2720–2733
67. Ding L, Wang Q, Shen M, Sun Y, Zhang X, Huang C, Chen J, Li R, Duan Y (2017) Thermoresponsive nanocomposite gel for local drug delivery to suppress the growth of glioma by inducing autophagy. Autophagy 13(7):1–15
68. Wang Y, Lin Y-X, Qiao Z-Y, An H-W, Qiao S-L, Wang L, Rajapaksha RPYJ, Wang H (2015) Self-assembled autophagy-inducing polymeric nanoparticles for breast cancer interference in-vivo. Adv Mater 27(16):2627–2634
69. Lin Y-X, Wang Y, Qiao S-L, An H-W, Zhang R-X, Qiao Z-Y, Rajapaksha RPYJ, Wang L, Wang H (2016) pH-sensitive polymeric nanoparticles modulate autophagic effect via lysosome impairment. Small 12(21):2921–2931
70. Lin Y-X, Gao Y-J, Wang Y, Qiao Z-Y, Fan G, Qiao S-L, Zhang R-X, Wang L, Wang H (2015) pH-sensitive polymeric nanoparticles with gold(i) compound payloads synergistically induce cancer cell death through modulation of autophagy. Mol Pharm 12(8):2869–2878

Printed in the United States
By Bookmasters